PENGUIN BOOKS

Barging Round Britain

ın Sergeant is a TV and radio journalist, and formerly the
ıC's Chief Political Correspondent. He remains a TV regular
shows such as *The One Show*, *Have I Got News For You* and *QI*.
was the star of Strictly Come Dancing in 2008, repeatedly
ishing bottom of the scoreboard but top of the public vote.

vid Bartley, in his life on dry land, is a writer and producer
television documentaries, specializing in arts and history
jects. He lives near the Grand Union Canal in west London.

Barging Round Britain

Exploring the history of our nation's canals and waterways

JOHN SERGEANT
and DAVID BARTLEY

PENGUIN BOOKS

PENGUIN BOOKS

UK | USA | Canada | Ireland | Australia
India | New Zealand | South Africa

Penguin Books is part of the Penguin Random House group of companies
whose addresses can be found at global.penguinrandomhouse.com.

First published by Michael Joseph 2015
Published in Penguin Books 2016

Text copyright © David Bartley and John Sergeant, 2015

See page 384 for photography and illustrations copyright

The moral right of the author and illustrator has been asserted

Text design by Claire Mason
Typeset in Garamond MT Std 12.5/14.75 pt by
Palimpsest Book Production Limited, Falkirk, Stirlingshire

Colour reproduction by TAG Publishing
Printed in Great Britain by Clays Ltd, St Ives plc

A CIP catalogue record for this book is available from the British Library

ISBN: 978-0-718-18064-5

Contents

Introduction

by John Sergeant

Oh the joys of television. I am in the backyard of the famous Wedgwood factory, and they have given me a pile of reject china on a trolley. All I have to do, with their full permission, is to behave like a bull in a china shop and tip the whole lot over. But how can I make it look natural and, of course, amusing? My first attempt fails. The trolley has slipped into a narrow ditch, but I cannot make it flip over. The ditch needs to be deeper. Next time it is perfect. The noise is ear-splitting, the shattered plates and remains of the jugs spin vigorously on the ground.

It may seem a daft way to learn anything, but the more I pushed this ungainly trolley, the more I was forced to think about the real difficulties faced by Josiah Wedgwood 200 years ago. Nowadays it is so straightforward. You go into a shop, have your plates packed up and off you go. Mr Wedgwood began by depending on horses and carts to deliver his delicate china. He longed to use canal boats. Then his goods could travel safely in huge quantities, to supply the British market and also satisfy the demands of eager buyers overseas. But his planned factory was a long way from a decent waterway, let alone the sea.

Mr Wedgwood was not the sort to give up. British manufacturing, spurred on by the industrial revolution, was leading the world. And he was convinced that his new style of incredibly thin bone china would make his name and his fortune. He dreamed of turning the Potteries into Eldorado. He would conquer the world, he announced, 'vase by vase'. Despite all the difficulties, he set off to build a waterway, which would become the famous Trent and Mersey Canal. Later a giant boat lift would be added, powered by steam engine. The resulting system sent clay in one direction and the finished china back in return, with almost no limit on how often it could be used. Engineering genius and a business wizard had produced the marvel that we can still see today.

Having been given the chance to travel along the best canals in Britain for an eight-part television series for ITV, I had a ringside seat to enjoy some of the most spectacular feats of civil engineering ever undertaken in this country. These beautiful canals were not designed to look good. They were born as a result of hard times and grim reality. I saw extraordinary examples of technical innovation, but I was also struck by the harsh requirements which drove the engineers to excel in such amazing ways. Even when the construction work was over, thousands of canal workers were needed to keep the whole system going through long hours and backbreaking toil. It was often the only way to keep their families from starvation.

Now, these pencil-like waterways, criss-crossing the country, are almost entirely devoted to holidays and the leisure industry. But they also remain a great historical treasure. Many people see them as one of the wonders of the world; and it is not hard to see why. They can be appreciated on so many different levels; and their charms are not always obvious. You can drive along a motorway, catch sight of a canal boat and not be unduly impressed. You may not relish the contrast between the speeding cars and the slow, deliberate pace of the narrow boats. But most of those on the water are unlikely to be in any doubt. They may well favour a 4 mph speed limit and believe, when it comes to lifestyle, they are the ones to be envied.

I discovered, though, that some of those who have spent years living on canal boats can still find it hard to explain why they would not want to base themselves on dry land. It is wonderfully complicated. There is an element of mystery. Most of the canal users certainly have a desire to escape from everyday concerns and they also like to be involved in some way with history, to live in the past. But why have the canals attracted so much more interest in recent years? More miles of the old waterways have been restored than ever before. Is it because we are so buffeted by technological change that we instinctively feel much closer to the original canal promoters, who were hit first by competition from the railways and then by the dramatic switch to the roads?

You may say this is all stuff and nonsense. A week or two on a canal, with bed and board included, is just a cheap family holiday, with Mum happy to be homemaker, Dad pretending he is a yardarm away from being Lord Nelson, and the children determined not to give up their role as twenty-first-century transformer fanatics. What I found, to my delight, is that people on canals are far more varied and interesting. There are the stay-on-boards, lots of them, who live on their boats, which they are proud to own. Some of them are reclusive, and a few of them live alone. But they share a desire to enjoy the boating life, moving on when they feel like it, choosing their neighbours and not being told where and how they should live.

There are the regular visitors who want to celebrate the countryside. They are taken by the wildlife, have no difficulty distinguishing a coot from a moorhen, are mesmerized by a hesitating heron and revel in the blue flash of a kingfisher. If you mention some of the less enticing aspects of the countryside, perhaps an attack of horseflies or locals giving a rather too graphic description of their favourite manure, they adopt the dreamy posture of the true convert. They have decided from the very beginning of the fortnight that this is going to be a perfect holiday and they will not be deflected.

There are the children, particularly those approaching their teens, who are confident they know how to get the best out of their time on the water. They are amused by those younger than themselves who don't know how to operate a lock or manage a swing bridge and when questioned admit, with a deep sense of satisfaction, that these accomplishments can also be difficult for grown-ups.

There are also those adults who, in spite of being confined to a narrow boat, want to make the most of this experience. They scan the maps, anxious to make sure they are always within walking distance of a good pub. With electronic equipment easily charged up on the boat when the engines are running, they can check on precisely the beers that will be on offer and, after looking over the sample menus, decide it may be wise to stick to the fish and chips. I have to admit that sometimes I fell into this category.

Then there are those I really admired, the canal volunteers drawn by a deep appreciation of the past and a determination to look after every aspect of the waterways. I joined a group in Bradford-on-Avon who had undertaken to clear the banks of undergrowth. They were a dedicated bunch, mainly retirees, who threw themselves into the task. One of the women pointed out that enthusiasm alone was not enough; we also had to make sure that the weeds I was cheerfully throwing into the canal were kept on board to be disposed of later. As the working boat rose through the muddy water of a lock, I asked whether our interest in the long-outdated canal system was yet another manifestation of the British character. One of the volunteers paused for thought as she stood in front of the dirty lock wall moving steadily upwards. 'Well,' she said finally, 'it is our heritage.'

I have thought about this question in some detail. There is certainly something very British in the way we cherish the canals, which for only a relatively short period provided the answer to our transport problems. Nostalgia is part of it, a feeling that we have left something valuable behind which can be retrieved; it takes us back to a time when the pace of life was slower and maybe suited us better. We also seem to be attracted to the idea that the way those undoubted hardships were overcome represents a peculiarly British kind of victory and should be celebrated. We also like to be reminded of that period in history after the industrial revolution when we led the world. Trade followed the flag, and the Union Jack fluttered across a quarter of the globe. It was the greatest empire the world has ever seen; and that is not a bad thought to contemplate as you sit at your mooring, sipping a pint of British beer.

Of course, we are not living anything like the life of nineteenth-century boatmen. If you go on a canal holiday you are simply playing a game compared with those rough and tough canal folk, who were often scorned by those in regular jobs and treated like Gipsies. The men and women employed on the narrow boats had to work all hours to see their cargoes arrived on time. If they had children old enough to help that might make things easier, but often it was a couple walking with the horse and taking turns on the tiller. I spoke to an elderly woman brought up on one of the few remaining working boats. And she described with bitterness her cruel childhood. She told how they had to work hard to earn enough to live; and they had to keep on the move. It could be bitterly cold in the winter and uncomfortably warm in the summer. However, despite all her hardships, she was still attached to the canals. This strange, shifting life outdoors was where she felt at home.

The more you know about the British canal system, the more it fascinates; the more you travel across this vast network, the more you realize how much you still have to learn. There are so many possible canal journeys, and they vary so much in character and type. From the great mountains and lochs which dominate the Caledonian Canal to the gently rolling countryside adorning the Kennet and Avon Canal there is so much to discover. The people I met ranged from hardy souls who had spent years on the canals to those who had just managed a few days. I can honestly say there were few cross words from us or anyone else. When the sun was shining people seemed to find it hard not to smile. This book contains descriptions of eight of what we regard as the best canal journeys in Britain, but it could very easily have been a longer list. One of the joys of boating is imagining the voyages you could take when you have the chance to shrug off the pressing concerns of everyday life. I hope this book will help.

There were many memorable moments as we worked our way across the country. I will not forget my encounter with Freddie, a gentle giant of a Shire horse, pulling narrow boats along a stretch of the canal at Kintbury in Berkshire. I was allowed to take his reins, and, with some gentle encouragement, and frequent stops to sample the grass, Freddie settled into his routine. It was a hot day, and I became increasingly unaware of anything except the task in hand. After about an hour, Freddie came to a halt. It was where he always stopped to allow the tourists to get off the boat, and I gave him his reward. A peppermint sucked across his large tongue narrowly missed his enormous teeth. His pleasure was obvious. I gave him another mint and then the rest of the packet. I had bonded with Freddie, and he had made my day. These programmes in the high summer of 2014 were among the most enjoyable of my entire career.

John Sergeant

Canals, from the 1760s until the 1840s, when they suffered a form of sclerosis thanks to the coming of the railways, were the arteries of the industrial revolution. It wasn't the case that this revolution came into being and then the canals helped it along. Neither could have developed without the existence of the other. Canals brought raw materials – coal, limestone and iron – to the infant industries that enabled the Age of Steam to roar into flame.

Inland towns without the benefit of navigable rivers now became behemoths of industry. The Potteries were able to source raw materials and ship their goods all over the world. The Black Country became a smoky industrial powerhouse, Birmingham the City of a Thousand Trades. The all-important textile industry either side of the Pennines was also well served by canals.

By the time the canal age got stuck in neutral, Britain had become the Workshop of the World, its products shipped to all four corners of the globe, its manufacturers and merchants rich beyond the dreams of avarice.

Goods could be carried up to four times more cheaply by canal than by road. Coal for domestic fuel, and later gas lighting, salt, soap and all manner of new, and cheaper, pots, pans, crockery, buttons, bows, bedsteads, trinkets and other creature comforts, transformed the lives of ordinary Britons. Materials to build the new Britain came by canal – bricks, stone and gravel. Thanks to canals the chemical industry picked up speed. Fertilizer was carried by water and helped foment what's been called the agrarian revolution, the striking improvements in agriculture, contemporary with those in industry.

But nothing lasts for ever. The glittering success of the first canals led to the 'Canal Mania' of the 1790s, when promoters, hucksters and get-rich-quick merchants whipped up a feeding frenzy of speculation in canal shares, tempting the unwary into schemes that seemed too good to be true and usually were. Many investors lost their shirts, as canals were left unfinished or saddled by crippling debts.

The railways, from 1830, put the biggest spoke in the wheel, and the canal companies, who'd been living high on the hog, panicked, sold up and sold out. Commercial trade on the canals survived until very recent times, but it was bleeding away from the First World War onwards, in a slow but steady haemorrhaging of traffic, revenue and hope.

Many were abandoned, and the whole network looked like slipping into the wilderness from which it had come – dry, rusting, rotting, moss-laden. But from the 1950s, in ever-increasing numbers, an army of dedicated volunteers began to reclaim these fallen giants, a process that continues today, bringing more and more miles of waterway, and the associated buildings that have survived, back into being. To be enjoyed by a competing alliance of pleasure boaters, canoeists, anglers, walkers, dog-walkers and dogs. Reservoirs are now chock-a-block with sailing and wind-surfing clubs, and the waterways themselves – thanks, ironically, to their near-abandonment – are havens of wildlife, where the rarest flora and fauna thrive.

Today there are more craft on our canals than there ever were during their industrial heyday. Leisure has succeeded in keeping canals open. That this should be the case would, surely, have astounded the men who brought them into being – as well as the women who played a vital part in keeping the boats and goods moving. Their story has only relatively recently begun to be told.

This book explores the history of British inland navigation by looking at each of the eight waterways travelled by John Sergeant in his ITV series *Barging Round Britain*: the Trent and Mersey Canal, Birmingham Canal Navigations, Aire and Calder Navigation, Leeds and Liverpool Canal, Grand Union Canal, Llangollen Canal, Caledonian Canal and Kennet and Avon Canal. We'll meet some of the most fantastical constructions of the canal age: the world-famous Pontcysyllte Aqueduct, on the Llangollen Canal; Caen Hill Lock Flight, on the Kennet and Avon; the Harecastle Tunnel and Anderton Boat Lift, on the Trent and Mersey; the Burnley Embankment and Bingley Five-Rise locks, both on the Leeds and Liverpool Canal; the notorious Blisworth Tunnel, on the Grand Union; and the extraordinary, gigantic Neptune's Staircase and Clachnaharry Sea Lock, the pride of the Caledonian Canal.

As well as their works, we'll look at the lives, talents and legacies of the great canal builders: John Smeaton (who coined the term civil engineer), James Brindley, William Jessop, Thomas Telford and John Rennie. And the contributions made by the titans of industry so intricately linked with the glories and success of the canal age – Josiah Wedgwood, Abraham Darby, Richard Arkwright, John 'Iron-Mad' Wilkinson, Matthew Boulton and James Watt.

①	Trent & Mersey
②	Birmingham Navigation
③	Aire & Calder
④	Leeds & Liverpool
⑤	Grand Union
⑥	Llangollen
⑦	Caledonian
⑧	Kennet & Avon

Britain's Canal Network

The Trent and Mersey Canal

Introduction by John Sergeant

When do you know you are in danger of becoming a real canal enthusiast? To me one of the answers to this important philosophical question can be summed up quite simply. It is when you are excited by the idea of a giant boat lift, which will effortlessly transport your craft, with all your crew and contents, from river to canal and back again; when you are ready, of course.

I was on the gently flowing River Weaver at the start of my journey along the Trent and Mersey Canal. Before 1875 I might well have been carrying salt from the local mines and be about to shift it up a steep slope to another boat on the canal itself. Coming the other way, I would have to slide my cargo – perhaps china from the Wedgwood works – down chutes to a boat on the River Weaver.

For more than a hundred years, initially using a steam engine powering a clever system of hydraulics, the Anderton Boat Lift could do it all for you. The trick was to pump water weighing about the same as a heavily laden canal boat into large tanks on one side of the iron structure. And on the other side the boats would be lifted gently upwards, as if by magic.

This is the last working boat lift of its kind in England. It was fully restored in 2002 and uses oil rather than water and electricity instead of steam. But the fun of making the crossing is unchanged. Skilled men using radios guide the boats through, and all you have to do is give a brief rendition of 'Rule Britannia'. It was the first of its kind in the world.

There are other first prizes which have been won by the Trent and Mersey Canal. It was a joint project mounted by Liverpool Corporation and the Staffordshire pottery owners led by Josiah Wedgwood. For the first time goods could travel by water from the Potteries to the port. And it involved building the first three British canal tunnels. It also deserves a first for its ingenious name. Yes, it does go along the River Trent, but it only briefly joins the river itself, and it never actually comes within sight of the River Mersey. But that is show business.

One of my most interesting experiences was to travel through the Harecastle Tunnel. It plunges through a hill, which would otherwise have required two flights of locks, and to begin with all was calm. The light from the entrance became a small white disc as we retreated into the darkness. This being television, my guide began to explain the legend of the local ghost; I felt less and less able to laugh it off when small pieces of masonry began falling off the roof. We spent a full forty minutes in the cold and damp before being gratefully released on the other side.

I was reminded of the old canal folk and how they had to cope. Usually the tunnels had no space for towpaths, so horses could not accompany them. Eventually steam tugs were made to pull the boats through. But before then the boats had to be 'legged'. Lying on their backs, the crew would push their way through as if they were walking along the roof of the tunnel. Professional 'leggers' would wait at the entrance in the hope of picking up work. As I gripped the handle to increase the power of our splendidly reliable engine, I gave a small, silent prayer to the twenty-first century.

The Anderton Boat Lift pictured in 1963

On 26 July 1766, at Middleport in the Staffordshire Potteries, a powdered and bewigged gentleman, thirty-six years of age, drove a spade into the ground, dug out a clod of earth and tossed it into a wheelbarrow. The man next to him then wheeled it away, to huge cheering from the large crowd that had gathered to watch this deceptively simple piece of business.

They were seeing history in the making. The man with the spade was Josiah Wedgwood, Father of English Pottery. The man pushing the wheelbarrow was James Brindley, the chief engineer and visionary of a project that was to change the fortunes, and appearance, of the whole country. This ceremonial cutting of the first sod of what was to become the Trent and Mersey Canal was the opening move in the creation of the 'Grand Cross', a network of inland waterways that would link up the four great English ports – Liverpool, Hull, Bristol and London – and through them tap into the vast wealth offered by world trade. The canal age had begun.

Wedgwood was the driving force behind the Grand Trunk Canal, as the Trent and Mersey was first known. What Henry Ford was to the economy of twentieth-century America, Josiah Wedgwood was to that of eighteenth-century Britain. Pottery, unlike the textile industries at the same time, didn't rely on mechanization.

But Wedgwood saw that it could be boosted from a small-scale 'domestic' industry into a global one. If only he had a canal.

Wedgwood was born in Burslem in 1730, the twelfth and youngest child of a potter. Like so many British industrialists, he came from a Non-conformist background. When he was nine his father died, and Josiah left school to become apprenticed to his older brother as a 'thrower'. But a childhood bout of smallpox had left his right leg fatally weakened. While he was highly adept at the art of throwing the clay on to the wheel, and moulding it into a pot, his leg was not strong enough to drive the treadle turning it. He turned his attention instead to the art of making glazes – and the science behind it. He became an obsessive experimenter in different glazing techniques.

Josiah Wedgwood (1730–95), the famous potter

English pottery at this stage was completely overshadowed by its foreign rivals, especially those at Delft, in the Netherlands. Wedgwood was to change that and make English pottery the world leader.

His first great breakthrough came with his improvement of a cream-coloured earthenware, to which he added a rich and brilliant glaze. Known originally as 'Useful Ware', later 'Cream Ware', it was a well-made product that the middle classes could afford. A green and gold service Wedgwood made for Queen Charlotte in 1765 so impressed her that he was subsequently allowed to call his product 'Queen's Ware' and was appointed potter to the Crown. The following year he began building his famous factory, Etruria (named to conjure up the pottery then thought to have been made by the ancient Etruscans), near Burslem.

Along with Queen's Ware, which became the leading type of pottery in Europe, Wedgwood also invented Jasper, an unglazed, delicately tinted product which has been called the most significant development in ceramics since

An 1866 engraving of the Wedgewood factory at Etruria

porcelain. He conducted over 5,000 experiments until he was satisfied with the results. Wedgwood also developed an important line in black basalt and invented the pyrometer, an important device used to measure oven temperatures for firing. This led to his being invited to join the Royal Society in 1783.

The second secret to Wedgwood's success was his genius – the word doesn't seem misplaced – as a businessman. An early brush with near-bankruptcy prompted him to become one of the pioneers of proper cost-accounting, keeping careful track of cash flow. He also realized the great benefit of efficient division of labour (famously advocated by another key figure of the British Enlightenment, the great Scottish economist Adam Smith, in *The Wealth of Nations*, published in 1776). Each part of the industrial process was broken down into stages and given to one set of workers to complete, making the whole, tightly organized operation quicker, cheaper and able to produce better-quality goods.

A man possessed of vast energy – despite having to have his right leg amputated when he was thirty-seven, thanks to the aftermath of the smallpox he had contracted as a child – Wedgwood personally supervised the output of his factories. Work that was in any way less than perfect he would smash to pieces with his walking stick, scrawling in chalk on the work bench: 'This won't do for Josiah Wedgwood.'

Unlike many later factory owners, Wedgwood took an enlightened interest in the well-being of his workers. When he constructed his estate, he built not only a great house for himself, Etruria Hall, but housing for his workers on the same site as well. Later he became a prominent supporter of the campaign to abolish slavery. His factory distributed, free of charge, a famous medallion depicting an enchained Negro slave, with its motto 'Am I Not a Man and a Brother?'

But no matter how brilliant Wedgwood's designs, how sparklingly efficient his management and accounting techniques, Etruria was not going to be worth the candle without effective transport. And this, up until the coming of the canal, the Potteries didn't have. There were no navigable rivers. And so, in the first place, china clay – local supplies not being good enough for Josiah Wedgwood in either quantity or quality – had to come from Cornwall, Dorset and Devon, making a laborious journey by sea to the Dee or Mersey estuary, and then up the navigable River Weaver to Winsford, whereupon it was transported by packhorse 20 or more miles to Wedgwood's factory. (With it came salt, used for glazing.) Flint, meanwhile, another vital raw material, came from the south coast round to Hull, up the navigable stretch of the River Trent and then to the Potteries by packhorse or wagon.

Roads in eighteenth-century Britain were notoriously useless. Dust traps in summer, quagmires in winter. Legend had it that potholes existed that could swallow a horse and rider whole. The Roman roads were still in use but were badly maintained. The techniques of building hard roads, like much other Roman technology (such as concrete), had been lost after they pulled the ripcord at the beginning of the fifth century, withdrew their remaining legions and abandoned the Britons to their fate.

Local parishes were put in charge of roads and naturally

saw little reason to expend time, money or energy on maintaining them for the benefit of passing 'foreigners'. (People earning under a certain annual wage were expected to work for six days a year for free on local roads or, if earning above that wage, lend carts or horses, or give money.)

The government introduced half-hearted attempts to redress the situation. Narrow wheels churned up the roads, so statutes were introduced to specify a minimum width. But wider wheels could carry carts with heavier loads, which made the roads worse, so the legislation was pointless.

Along with the public highways were 'drove' roads, where animals were driven, often along huge distances, to market (or slaughter). There were also narrow 'pack roads', on which horses carried as much as they could in panniers. Highwaymen, far less romantic in real life than in later fiction, plagued the roads.

Thomas Bentley (1731–80), the porcelain manufacturer and business partner of Wedgewood

As the eighteenth century progressed the highways did begin to improve – in some cases at least. This was thanks to the introduction of turnpikes, or toll roads, maintained by 'turnpike trusts'. But the turnpikes still lacked a good hard surface – that would only come later.

The agricultural reformer Arthur Young, who travelled the country – as best he could – to inspect farms, said of the turnpikes of Lancashire, in 1770,

> *It is impossible to describe these infernal roads in terms adequate to their deserts . . . Let me most seriously caution all travellers who may accidentally purpose to travel this terrible country, to avoid it as they would the devil; for a thousand to one but they break their necks or their limbs by overthrows or breakings down. They will here meet with rutts which I actually measured four feet deep, and floating with mud only from a wet summer; what therefore must it be after a winter?*

You didn't need to be a business genius to see that moving fragile goods like pottery over pitted and pocked roads, in panniers that smashed the pieces into smithereens anyway, was not a sound business model. Wedgwood reckoned on two-thirds of his pottery being broken in transport.

And so, in 1765, Wedgwood, along with his friend and future business partner Thomas Bentley of Liverpool, and Erasmus Darwin of Lichfield, began to promote the idea of a canal from the Trent, east of Burton, to Liverpool. (Erasmus Darwin, a poet, physician, scientist and philosopher, was one of the grandfathers of Charles Darwin. Josiah Wedgwood was the other.)

Liverpool merchants signed up to the scheme, and so did Wedgwood's fellow potters.

This was not the first such scheme proposed. In 1758 the first arm of the wondrous cross had already been surveyed. A plan to build a canal 'with Locks to pound the water and make it dead as in Holland', from Stoke-on-Trent to Wilden Ferry, from where the Trent was navigable down to the Humber estuary, was being promoted by a man who was, along with Wedgwood, to become one of the most important players in early canal history: Granville Leveson-Gower, later 1st Marquess of Stafford.

Earl, or Lord, Gower was a substantial landowner and a considerable operator in the politics of his day, taking high office as well as prestigious royal appointments. At the time of the survey he was Master of the Horse. He was also MP for Lichfield, and Lord Lieutenant of Staffordshire. With his land, wealth and political clout – at local as well as national level – he was to be crucial to the eventual success of the Trent and Mersey Canal.

The man Earl Gower chose to carry out his survey was James Brindley. The man with the wheelbarrow who had joined Wedgwood in the ceremonial cutting of the first sod of the Trent and Mersey. Brindley did more than anyone to bring the canal age into being.

The two men, earl and engineer, could hardly have been more different. Brindley was born, in 1716, into genteel rural poverty in Tunstead, Derbyshire. His feckless father was the black sheep of a local Quaker family. Brindley received no formal education whatever, except what his mother taught him at home. In his later journals, or 'day-books', his spelling is eccentric even by the standards of the eighteenth century. He writes of performing an 'ochilor servey' or 'ricconitoring'.

He began his working life as a farm labourer. At the age of seventeen he was apprenticed to a mill- and wheelwright in Macclesfield. Here – and he seems to have had a photographic memory – Brindley's innate genius for mechanics first showed itself, and his reputation for problem-solving grew. Having opened his own mill-wright business, in Leek, Staffordshire, after the death of his master, he began to be invited to consult on any engineering conundrum, going further and further afield.

He invented an ingenious method of pumping water from mines in a colliery near Manchester and designed a water-power system for a silk mill in Congleton. He was also well acquainted with steam engines. As a mill-wright he was used to the ways of water, and construction in stone, brick and timber. If an apprenticeship for canal building had been designed, before such a profession had actually come into being, it might have looked like this.

Erasmus Darwin called Brindley a 'workman genius'. He was also a man of eccentric habits. His colleague and future brother-in-law, the surveyor and engineer Hugh Henshall, said of him:

When any extraordinary difficulty occurred to Mr Brindley in the execution of his works, having little or no assistance from books or the labours of other men, his resources lay within himself. In order, therefore, to be quiet and uninterrupted whilst he was in search of the necessary expedients, he generally retired to his bed; and he has been known to be there one, two, or three days, till he had attained

the object in view. He would then get up and execute his design,
without any drawing or model.

Brindley surveyed a route from Stoke to Wilden Ferry, and in 1761 it was checked, and approved, by the other great engineer of the early canal age, John Smeaton, who was in charge of the Calder and Hebble Navigation works. Smeaton added an intriguing suggestion. Could not the canal be extended 'to join the navigable river that falls into the west sea'? (He was presumably referring to the Weaver, which joins the Mersey estuary near Frodsham, in Cheshire.)

The same thought, independently, had occurred to James Brindley. And he went one better. Could not the canal connecting the Mersey and Humber estuaries be intersected by others, which linked to the Severn and Thames? The 'Grand Cross' was taking shape.

In 1759, the year after he had surveyed the Stoke to Wilden Ferry section, Brindley set to work on what became known as Britain's first canal. It wasn't actually Britain's first canal, as it happens, but it was certainly the most important. It was built at the behest of Earl Gower's brother-in-law, the 3rd Duke of Bridgewater. The man who would become known as the 'Canal Duke' was, along with James Brindley, one of the founding fathers of British canals. He constructed, entirely at his own expense, what would become the western arm of the Trent and Mersey Canal.

Born in 1736, Francis Egerton was a sickly child who wasn't expected to survive into adulthood. At the age of twelve he inherited the title after the death of his older brother. His mother had neglected to secure an education

The 3rd Duke of
Bridgewater
(1736–1803), creator
of the Bridgewater
Canal

for the boy because she thought he was an imbecile. He
was sent to Eton, at the behest of his guardians (one of
whom was his future brother-in-law, Earl Gower). How
he got on there isn't recorded. We do know that he went,
like other young male aristocrats of the day, on the Grand
Tour, a cultural odyssey (or so their parents hoped) which
could last up to three years. A tutor was engaged to show
his charges around the relics of the ancient world and the
glories of the Renaissance. Along with Italy, France was
also on the itinerary, war permitting. It is possible, some
think certain, that here Bridgewater encountered the
famous Canal du Midi, which had been constructed in
1681, as part of an ambitious plan to link the Atlantic at
Bordeaux to the Mediterranean at Toulouse.

But the spur to the Duke of Bridgewater's involvement

in his own canal project was that he was unlucky in love. Or in the marriage market, at least. He had sought to become engaged to one of the great beauties of the fashionable London scene, Elizabeth Gunning, the widowed Duchess of Hamilton. A squabble over the scandalous behaviour of her sister – a countess – led to the unhappy couple falling out. The engagement was off. (After this the duke shunned and abhorred the company of women, even to the point of refusing to employ female servants.)

He retired to his estate in Worsley, near Manchester, in high dudgeon. Here he took solace by turning his attention to his business interests. For he, like Earl Gower, was very far from being a mere pleasure-seeking drone. While the aristocracy's wealth, prestige and power were principally derived from land, with its rents and farm produce, Palladian country houses and biddable local MPs, many lucky landowners had discovered that underneath their fields was the most crucial commodity of the industrial revolution. Coal.

Coal

The 'black diamond' was the main driver of the great leap forward in industry, an important breakthrough in which came in 1708, when Abraham Darby first successfully smelted iron ore using coke (roasted coal) to make pig iron (so-called because the small troughs into which the iron was poured resembled suckling piglets) in a blast furnace. Previously iron ore had been smelted by using charcoal. But supplies of timber were running out, and from Elizabethan times coal began to take its place. Darby's factory was at Coalbrookdale in Shropshire – its name giving a clue as to the amounts of fuel to be had there.

Coke's ability to generate greater heat than charcoal led to advances being made in getting rid of the impurities in iron ore and facilitated the making of cast iron (less brittle and more malleable than pig iron), and later wrought iron (more malleable still). Used in the making of machinery (and subsequently machine tools), rails, bridges, roads, locomotives and boats, iron would transform Britain.

Coal's other crucial industrial role was to generate steam power. Thomas Newcomen's steam engine was first used to practical effect, in 1712, to pump water out of mines. In 1742 a Newcomen engine was installed at the blast furnace at Coalbrookdale. As the century wound on, an insatiable demand for steam power, and thus coal, drove the industrial revolution on. This in turn stimulated the demand for canals.

No wonder that the Duke of Bridgewater now began to take a close interest in his mines at Worsley. And in his mine-workers' time-keeping habits. He was an early advocate of the twenty-four-hour clock. Noticing that his workers had failed to appear on time after their dinner hour, he asked why not and was told that they had failed to hear the clock strike one, the signal for their return. The duke ordered that the clock should be altered so that it would strike thirteen. Unluckily for his workers.

Like Wedgwood, the duke had a problem – effective transport. Or rather, the lack of it. Although his mines at Worsley were only seven miles west of Manchester, as the crow flies, in transport terms they were a world away. And Manchester needed coal, as it was already expanding into the world's leading manufacturer of cotton. There were no coal mines in Manchester – it had to be brought in from outside.

And here the duke's land agent, or business manager, John Gilbert, comes into the canal story. (And was fairly quickly written out of it. Only recently has the vital importance of Gilbert been rediscovered.) Gilbert was an accomplished mining engineer and surveyor as well as all-round businessman. Scions of a long line of yeomen, he and his older brother Thomas (who worked for Earl Gower in the same capacity) were prototypes of eighteenth-century minor gentry capitalists, with their fingers in many pies, adept at several interlinked practical skills. He it was who now made a decisive intervention and pressed his employer to consider water transport.

So unprecedented was the waterway that resulted – the Bridgewater Canal – and so influential was it when news of its construction emerged that it has correctly been seen as the catalyst for the canal age. But canals *per se* were nothing new.

Canals: the Pre-history

They were known across the ancient world, from Meso-potamia to India, Egypt to Greece. The Chinese, as well as inventing the blast furnace, cast iron and steel (and gunpowder, paper, printing and porcelain), were also masters of canal work. They pioneered the use of 'pound-locks' and were known to have dug cuttings 80 feet deep. The seventh-century AD 'Grand Canal' of China is certainly that. Still the world's longest canal, it's a leviathan up to 60 yards wide and 1,115 miles in length.

In Britain the Romans built waterways for the transport of military materials. They also built them to drain marshland, as excavations in Car Dyke in the Fens have shown. In use from around AD 50, the canal found here was big – more than 40 feet wide. Foss Dyke, connecting Lincoln to the Trent, is still in operation.

In medieval times canals were dug to provide water for mills and ferry stone from quarries, and were often constructed at the behest of monasteries. Substantial canals were dug by Dutch engineers to drain the Fens in the seventeenth century.

Canals had also been made so as to make rivers more navigable, beginning at the River Exe, in Devon. But these early canals essentially 'improved' rivers (for transport purposes at least). They cut through meandering bends but they paralleled the river's course, giving them the great advantage of sharing its drainage basin. A 'true canal' crosses over a drainage divide, thus joining two separate drainage basins. (The first canal in Europe to connect two river valleys was the Canal de Briare, in France, completed in 1642.)

Although the Bridgewater is often referred to as the first 'true canal', it was actually preceded by the Newry Canal in Ulster, opened in 1742. This was built by Thomas Steers, the chief engineer of Liverpool Docks. At 18 miles long, 45 feet wide and more than 5 feet deep, this was a canal for sea-going ships. But by the 1780s problems with maintenance and flaws in its construction had led to its decline as a waterway, one reason why it has faded from the record.

In 1757 a pupil of Steers, and his successor at Liverpool Docks as chief engineer, Henry Berry, built the Sankey Canal. This was a more direct model for the Bridgewater and much closer to home. It carried coal from St Helens to Liverpool, where it was used in salt works, while salt made the journey back to industrial south Lancashire. (It's sometimes called the St Helens Canal.) The reason why it was missed off the roll of honour is that it was made in secret.

The Sankey Brook Navigation, as it was officially called, was sold to parliament as just that, a navigation, or 'improvement' to a river. But, in reality, it was to be a canal. Berry and the principal shareholder agreed that its true status was to be concealed – from the other proprietors and from parliament – on the grounds that such a novel undertaking would be met with opposition. Instead, the enterprise was passed off as an extreme version of a navigation, a navigation plus, even though the majority of its length was canalized.

Whether or not the Sankey, which closely shadows an existing waterway, is a 'true' canal, and thus the first on the mainland, or a navigation, is still debated. Certainly, at 8 miles long, with ten locks, it impressed the Duke of Bridgewater. Samuel Simcock, one of Brindley's assistants, was packed off to take a close look at it.

The duke liked what he saw. In 1759 he decided to cross the Rubicon and build his own canal. An Act of Parliament was obtained, and work could begin.

One thing the duke had perhaps learned from the Sankey Canal was the value of keeping your cards close to your chest where canal building was concerned. The first Act of 1759 sanctioned not one but two canals, one going from Worsley to Salford, keeping all the while to the north of the River Irwell, another to join the Mersey at Hollins Ferry. From here traffic could be carried into Manchester. The proprietors of the Mersey and Irwell Navigation thus raised no objections to the scheme.

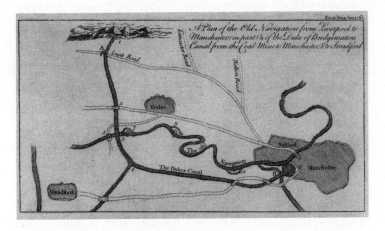

1766 map of the proposed canal route between Manchester, Worsley and the Mersey

But the canal to Hollins Ferry, though a couple of miles of it was built, may have been just a ruse, a diversionary tactic. According to this theory, the duke intended all along to take his canal into the heart of Manchester, cutting the Mersey and Irwell Navigation out of the deal altogether. (An alternative idea is that he changed his mind about joining the river at Hollins Ferry because the Mersey and

Irwell proprietors were demanding extortionate tolls.) Either way, the duke had thought again. A few months after work had begun, he dramatically moved the goalposts. From parliament he sought and obtained permission to change his route into Manchester.

The new plan was an astoundingly bold one. Not only would the Bridgewater become the first British canal that didn't join up with a river. Not only would it cross a river's watershed. The 'Duke's Cut', as it was at first known, was going to cross the river itself. The men charged with effecting this unprecedented piece of engineering were John Gilbert and James Brindley.

The scheme unfolded to howls of pain and fury from the proprietors of the Mersey and Irwell Navigation. But there was little they could do, because the earlier Act had already set the precedent for the Bridgewater Canal to reach Manchester. Their objections to the fact that they would suffer the indignity of the rival waterway crossing over theirs, thumbing its nose at them as it did so, were too late.

What was particularly galling for them was that the company had actually obtained an Act of Parliament as far back as 1737, allowing them to make the Worsley Brook navigable, thus doing essentially the same job as the Duke was doing now. The Mersey and Irwell proprietors were to repent at leisure for their predecessors' inertia.

The aqueduct crossing the Irwell at Barton was certainly not the first built in Britain – again the Romans got there first – but it was the first British aqueduct to carry a working waterway, boats and all. The Canal du Midi, which His Grace may have visited on his travels, also did so. (An 'eminent engineer' consulted by the duke, name unknown, scoffed that 'I have often heard of castles in the air but never before have I seen where any of them were to be erected.' Clearly he hadn't been on the Grand Tour.)

The duke kept faith, and John Gilbert was put in overall charge of building the canal, the aqueduct and the 200 yard embankment at Barton needed to raise it to the required height. James Brindley, as chief engineer, got the job of making the canal actually work.

The Bridgewater Canal passes over Barton Bridge

In many ways Brindley was setting out from scratch. Although as a former mill-wright he had a good grounding in water-management, he had, pretty much, to make things up as he went along. In doing so he, and the so-called

'School of Engineers' who trained under him, laid down the template followed by all the canal-builders who came after.

Brindley's first problem was how to stop the canal leaking water. His solution was 'puddling', a mixture of clay and water with which he lined his canals. (The Romans had, of course, already invented it. But Brindley was working in the dark.) The puddling technique proved up to the job and was copied in future canals.

As the Barton Aqueduct neared completion, there was feverish speculation as to whether the whole thing would simply collapse on its debut. (There had already been slips along the way.) Excitement, trepidation and, if there were members of the Mersey and Irwell company present, a happy anticipation of *schadenfreude* buzzed through the large crowd of onlookers who gathered at Barton, on 17 July 1761, to watch the first consignment of boats cross over the river. Or fall into it.

Anyone hoping for the latter outcome was to be disappointed. The boats passed without mishap. The *Manchester Mercury* reported that 'it is with pleasure that we can inform the Public that the experience answered the most sanguine Expectations of everyone present'. The Duke's Cut was a triumph.

The whole operation was equally efficient. At Worsley Delph, John Gilbert and Brindley were to dig an astounding 46 miles of tunnels into the mine itself. The tunnels helped provide the all-important drainage, as well as a water source for the canal.

From the hive of tunnels coal could be loaded straight on to special boats designed for the purpose. These were eccentric craft nicknamed 'starvationers' because of the skeletal appearance of their exposed ribs. This double-ended boat, with a rudder that could be swapped from one end to the other, took coal from the coal mines directly. It's believed that the starvationer was the direct ancestor of the 'narrow boat' (which was also, initially, designed to have a detachable rudder so as to be able to move in either direction without the need for 'winding', or turning around).

The starvationers were an early, if very small, form of 'containerization'. Specially designed containers of coal were loaded on to the boats, and then five or six would be tied together and pulled by horse-power down to the duke's own wharf and warehouses in Castlefield, Manchester. Here the containers were unloaded. The system worked. The price of coal in Manchester fell by half, from 7d a ton to 3½d.

The next year, 1762, the duke, Gilbert and Brindley began building an extension to the Bridgewater Canal, running from Trafford Park ('Waters Meeting') in west Manchester towards the Hempstones, on the Mersey estuary near Halton. The Act enabling it was passed despite frantic opposition from the Mersey and Irwell Navigation Company, now alive to the scale of their neighbour's ambitions.

Thanks to His Grace, the Age of Canals had been inaugurated. For seventy years they were to reign supreme.

Josiah Wedgwood wasn't present at the first crossing of the Barton Aqueduct. Nor was James Brindley. He had, as was his habit on troublesome occasions, retired to his bed. But Wedgwood, of course, knew all about the

Bridgewater Canal and pressed ahead with his own. James Brindley was hired to survey and engineer it. (Unlike John Gilbert, who was the duke's man, Brindley was freelance.)

The original plan that Wedgwood, Brindley and their supporters favoured was to join their canal's western end to the River Weaver at Frodsham Bridge, and thence to the Mersey estuary. Significantly, this was just below the jurisdiction of the proprietors of the Weaver Navigation, who regarded the proposed new arrival with horror and indignation. The matter seemed to be settled, and Erasmus Darwin began enthusiastically circulating pamphlets endorsing the scheme. Bold plans to extend it to both the Severn and Thames were also excitedly discussed.

But then in April 1765 an even more ambitious plan was proposed. Wedgwood had a meeting with John Gilbert, and naturally they discussed the canal. Why not, Gilbert suggested, go one better and link the new canal not just with the Mersey estuary but with the Bridgewater Canal, and thus Manchester too? Gilbert outlined a bold scheme. The extension of the Bridgewater, instead of aiming for Hempstones, could make for the area around Preston Brook. Here it could join the Trent and Mersey, and the duke, at his own expense, could construct a line down to the estuary itself. Wedgwood was tempted.

The duke himself was enthusiastic, and so too were Earl Gower and Thomas Gilbert, the latter's agent. Gilbert suggested that a PR campaign should begin. Wedgwood promptly wrote to his friend Thomas Bentley in Liverpool, asking him 'to draw his quill for the service of his Country' and write a pamphlet to explain the canal's benefit to the public good. The missive was titled *A View of the Advantages*

of Inland Navigation, with a plan of a Navigable Canal intended
for a Communication between the Ports of Liverpool and Hull.

The Weaver Navigation proprietors, meanwhile, realized that the prize of trade with the Potteries was slipping out of their grasp. Their route, after all, was the one by which the all-important china clay came by. The Weaver also shipped coal, and salt, which was needed for ceramic glazes.

The proprietors now proposed a canal of their own. This, naturally, would employ the whole of the Weaver Navigation, down to Winsford, and then make its way to the Trent. They solicited Wedgwood and his colleagues to sign up to the new scheme. The Weaver interest wanted a partner, not a rival. Wedgwood and his fellow potters were tempted by this idea too.

Here was an early example of the power politics that so often surrounded the formation of canal schemes, with interested parties (never mind those who objected) vying for leverage and jockeying for position behind the scenes.

Another issue still to be solved was where the eastern end of the canal would join the Trent. The river was navigable from the Humber estuary up to Burton-on-Trent. Here, the proprietors of the Trent Navigation, like their counterparts on the Weaver, were dismayed at finding they were going to have a cuckoo in the nest.

Until the coming of canals, a river navigation was, of course, the only game in town when it came to water transport. The proprietors – or 'undertakers' – of such schemes naturally took full advantage of their monopolies. One great advantage of the canal age was that it offered competition to these overpriced concerns.

The Trent Navigation had been authorized in 1699 and

gave one Earl Paget, who owned land and coal in the region, the sole right to own or approve all wharves from Nottingham to Burton. The people of Nottingham had, unsurprisingly, fiercely opposed the legislation.

Paget then leased the navigation rights to the only two men who owned wharves, at Burton and Wilden Ferry respectively. They were also boat owners and insisted that only their boats could use their wharves. Thus they controlled the carrying trade on the navigable Trent, cutting off Nottingham's route to the Potteries and to the Humber, unless its merchants paid through the nose for the privilege of using their facilities. And with a captive market, little need was felt to make improvements to the navigation itself, which flooded in winter and dried up in summer.

When merchants from Nottingham tried to circumvent this stranglehold and unload goods at other points on the river where they could be carried off by carts, the owner of the wharf at Wilden Ferry placed a string of barges across the river, which were vigorously defended. His counterpart at Burton sank a large barge just outside a lock, forcing goods to be trans-shipped around it.

The Trent and Mersey Company had no intention of being held to ransom by such an outfit, so they pressed on with their plans and got parliamentary permission to build their canal to join the Trent downstream of Wilden Ferry, past the point where Paget's writ ran. They set the compass for Derwent Mouth, where the River Derwent flows from Derby to join the Trent. Shardlow, near the confluence of river and canal, became an important inland port.

Boats could then make their way down to Trent Falls, where the river meets the Yorkshire Ouse at the Humber estuary. From Wilden Ferry to the estuary there were no proprietors of navigations to worry about – passage on the river was free.

The glory days for the greedy owners of the Burton Navigation were soon to pass – the canal more or less paralleled the Trent from Burton to Wilden Ferry. The eastern route decided, the main protagonists met up at an inn in Wolseley Bridge, near Rugeley, in December 1765. Earl Gower was in the chair, and his agent Thomas Gilbert (MP for Newcastle-under-Lyme) was present, as well as his fellow local MPs. James Brindley was on hand to outline the scheme. As Wedgwood reported, 'Brindley was called upon to state his plans, [and] brought them forward with such extraordinary lucidity of detail as to make them clear to the dullest intellect present.'

Samuel Smiles, the Victorian promoter of 'Self Help', and one of the first substantial historians of canals, takes up the story in his *Lives of the Engineers*.

The promoters of the measure proposed to designate the undertaking 'The Canal from the Trent to the Mersey;' but Brindley, with sagacious foresight, urged that it should be called The Grand Trunk, because, in his judgment, numerous other canals would branch out from it at various points of its course, in like manner as the arteries of the human system branch out from the aorta; and before many years had passed, his anticipations in this respect were fully realized.

All was agreed, and money was solicited to put the scheme in action. Wedgwood himself promised £1,000 towards the preliminary expenses.

That the principals believed in the public good the canal

would offer there is no reason to doubt. The motto of the Trent and Mersey was 'Pro patriam populumque fluit' ('It flows for country and people'). Like other early canals, it wasn't set up with the principal aim of making profits from tolls. Its inspiration was to facilitate transport, and thus trade.

The business set-up initially favoured by Wedgwood and others amongst his colleagues was that, like the turn-pikes, the canal should operate as a trust. Thus, though it would charge tolls, it would use its income for the upkeep of the canal, rather than make a profit for its sharehold-ers. The Liverpool merchants were keen for it to become a company, though, and cited the example of the Sankey Canal as one such that had nonetheless provided much public benefit. Had the Trent and Mersey become a trust, which would have meant that ultimately tolls would have disappeared, as they did on the public roads, the future of the canal system may well have been different.

But Wedgwood, though he had said of the canal that 'the very term *Private Property* is obnoxious', came to feel that a trust would be unworkable, requiring too much time of public-spirited individuals working only *pro bono publico*.

Thus the Trent and Mersey was formed as a joint stock company. Such outfits had been out of favour ever since the investment scheme known as the South Sea Bubble had spectacularly popped in 1720, leaving thousands of inves-tors and speculators flat bust (including Sir Isaac Newton, who remarked mournfully that 'I can calculate the move-ment of the stars, but not the madness of men'). The resulting scandal (members of the government were heavily invested) had meant that further joint stock companies could only be formed if an Act of Parliament granting a royal charter was passed in their favour. In order to become joint stock companies (and they needed to raise vast amounts of capital, making this approach the most practical), canal interests had to demonstrate their financial viability and probity, as well as the public benefits of their waterways, in order to have any chance of going ahead.

Although the Trent and Mersey Company's financial structure was decided on, its western terminal hung in the balance. It became clear to Wedgwood and his supporters that the Bridgewater–Gower nexus was intent on wresting control of this section of the canal away from the potters and Liverpool merchants and into their own grasp. What's more, the duke was worryingly vague as to where precisely his proposed extension would join the Mersey. Thomas Bentley and his Liverpool allies were pressing him to build an aqueduct over the estuary to the Lancashire side, thus linking more directly with the port itself. Whether the duke was concealing his plans, or simply didn't know what they were, is not clear. But Bentley began to suspect that the Wedgwood–Liverpool faction was being played for chumps.

The Weaver proprietors, meanwhile, were putting pressure on the Wedgwood alliance by proposing a canal of their own, which would run from Winsford to the Trent by passing north of Stafford – thus missing out the Potteries altogether. Wedgwood and his fellow potters were not over-keen on such a plan. And yet they were worried by it. 'Though I believe that they meant it first merely as a *Bug Bear* to us I do not think,' Wedgwood wrote to Bentley, 'we should treat it altogether as a chimerical scheme.'

Wedgwood now made his mind up. He decided that if

Gower – who was lord lieutenant of Staffordshire, after all – was seen to be playing fast and loose where the interests of the Potteries were concerned he would never be able to show his face in the county again. The die was cast. The Trent and Mersey would join the Bridgewater Canal extension at Preston Brook. The duke would then build the final spur, which would fall into the Mersey at some as yet unspecified point near Runcorn. There would be no aqueduct across the Mersey. It's doubtful the duke had ever seriously intended for there to be one. He now controlled the western terminus of the Trent and Mersey Canal and its trade with both Liverpool and Manchester.

But first the canal had to get an Authorizing Act of Parliament. And with so many vested interests ranged against them, the promoters knew that they would have a battle on their hands to get the canal agreed to. The process experienced by the Trent and Mersey – and it could hardly, like the Sankey or Bridgewater canals, be executed by means of subterfuge – set the pattern for all canal companies to come.

A public meeting would be called. Subscriptions would be solicited to pay for the initial costs. (Such early subscribers would get the first shares, should an Act be passed.) An initial survey was done, and costs and revenues estimated. If the figures seemed to stack up, a Bill would be presented to parliament by the provisional management committee.

In support of the Bill a full survey had to be carried out, and a map of the proposed route drawn and submitted for scrutiny. The paperwork needed to indicate what land had to be bought, what rivers or streams needed to be altered, how water would be supplied and how much used, along with a detailed estimate of costs and likely revenues. (The preconditions of presenting a Bill would become more stringent later in the century.)

To keep the project financed through the three-stage process – two readings of the Bill followed by committees in both Houses of Parliament – required considerable financial outlay without certainty of reward. Canals were in every sense a major capitalist development.

A canal company, once authorized by parliament, had powers allowing for the compulsory purchase of the lands through which it passed. Any landowner opposed to his land being carved up by a canal thus made representations against the Bill. Objections were also made by vested interests affected commercially by the canal's progress. Thus a Bill was often opposed by a vituperative and devious alliance of landowners, millers – worried about what the canal might do to their water supply – and, of course, the owners of rival waterways, or of other means of transport that might be adversely affected. Plus rivals in the trades the canal was designed to foster. Such adversaries would often succeed in getting a Bill thrown out, or the route of a canal radically altered. (There was, of course, much horse-trading behind the scenes.)

The surveyor might be called on to answer questions at a parliamentary committee. There are amusing tales of Brindley going to Westminster, demonstrating the way to make clay 'puddling', chalking out designs on the floor and building working models made of cheese for the instruction of his inquisitors.

It became the pattern for each side to publish pamphlets,

arguing, in forthright and increasingly bad-tempered terms, the merits and demerits of the canal proposed, often highly exaggerated and fanciful in each direction. Palms would be greased, stories spun, journalists briefed, and, most crucially of all, members of both Houses of Parliament quietly nobbled.

In the case of the Trent and Mersey Canal, the Weaver and Burton Navigations were opposed, and so too was the Duke of Bridgewater's old adversary, the Mersey and Irwell Navigation. (Pamphlets published by them were loftily dismissed by the Trent and Mersey's own spinners as 'exposing to public View the flimsy Pretexts, under which the Enemies of Justice are driven to shelter themselves'.)

The supporters of canals constantly stressed the philanthropic and social benefits they would bring – especially, of course, to the very poorest members of the community. In 1769 a pamphleteer tried to appeal to the better nature of those objecting to the proposed Coventry and Oxford Canals.

I beg to ask the opposers of the navigation from Coventry to Oxford, whether they think all the inconveniences they have pointed out, are equal to the starving to death of many hundreds of poor inhabitants in the inland counties, where coals may be brought by such a navigation? There are many places near Banbury, Brackley, Bicester etc where coals cannot possibly be had at any price to supply the necessary demand for them . . . there are no materials by which the poor can procure even a wretched fire. In these parts, I believe, they often perish by cold . . .

It wasn't just landowners or commercial rivals that the canal promoters had to contend with. Samuel Smiles identified an equally intractable body of opinion.

There was also a strong party opposed to all canals whatever – the party of croakers, who are always found in opposition to improved communications, whether in the shape of turnpike roads, canals, or railways. They prophesied that if the proposed canals were made, the country would be ruined, the breed of English horses would be destroyed, the innkeepers would be made bankrupts, and the pack-horses and their drivers would be deprived of their subsistence. It was even said that the canals, by putting a stop to the coasting trade, would destroy the race of seamen.

As far as the Trent and Mersey was concerned, Wedgwood, with the support of his fellow businessmen, and the local MPs, had a strong hand. But his ace was the lobbying power of Earl Gower. Not only could he whisper honeyed words into the ears of his fellow lord-ships in the Upper House. The fact that his agent, Thomas Gilbert, as MP for Newcastle-under-Lyme, chaired the parliamentary committee to which the Bill was referred after its second reading helped too.

The influence of Gower and Gilbert was decisive. An Act of Authorization was passed in May 1766. The Trent and Mersey Company was in business. The building of the canal could begin.

Shareholders included Earl Gower, the Duke of Bridge-water, Thomas Gilbert, James Brindley and his brother John. Initially neither Wedgwood, Bentley nor Darwin

47

were shareholders, presumably to demonstrate their disinterested public-spiritedness.

The company officers, as Wedgwood proudly revealed to Bentley, were to be remunerated as follows:

James Brindley – £200 per annum
Hugh Henshall, Clerk of Works – £150 (to include cost of under clerk) p.a.
Josiah Wedgwood, Treasurer – £000.00 p.a.

(Later canal treasurers were somewhat better remunerated. It was, after all, a highly responsible and often fractious job.)

The biggest problem the company had to face was how to cross the ridge of the Peak District, the southern spur of the Pennines. The narrowest part of the ridge was found to be at Harecastle, near Kidsgrove, and so Brindley resolved to build a tunnel here.

Work began and the section from Wilden Ferry towards Harecastle went reasonably smoothly. Two aqueducts were built, over the River Dove near Burton and over the Trent ('Brindley's Bank') near Rugeley. A 130-yard tunnel was built at Armitage (subsequently home to Armitage Shanks, famous for their toilets and other bathroom appliances).

By November 1771 the canal had reached Stone, in Staffordshire, which would become the company HQ. At the celebrations which followed, cannons were repeatedly fired, and as a result a lock and bridge collapsed. They cost £1,000 to repair. Undaunted, the canal reached Stoke-on-Trent in September 1772.

The section falling from the Harecastle summit to the Cheshire Plains down to Middlewich, with its thirty-five

'Cheshire Locks', was slow going, however, only being completed in 1775. From Middlewich to Acton Bridge the route was even more intractable, thanks to boggy ground, and the canal diggings slipped.

The biggest hold-up of all was at Harecastle. Although, of course, tunnels were hardly unknown, Brindley was again working in the dark – in more ways than one. This was to be the first canal tunnel in the country.

Brindley's plan at Harecastle was similar to what he and Gilbert had done at Worsley. The tunnels themselves would provide the water for the canal. Coal, they hoped, would be dug out as an added bonus. Shafts were sunk down to the level of the proposed canal, and earth drawn out by horse-gins, with windmills and watermills also working pumps. When great quantities of water were encountered, Brindley's previous knowledge of steam engines for pumping came to the fore. He designed one especially for the tunnel workings.

Entrance to the first Harecastle Tunnel, built by James Brindley

The Harecastle Tunnel would be 2,880 yards long and 12 feet high. To save time and money, it would be 9 feet wide. This was to be a fateful decision. Only narrow boats would be able to make a through journey on the Trent and Mersey Canal.

What Brindley hadn't realized, until the work got going, was how extremely hard the rock through which he aimed to tunnel actually was. (When Thomas Telford came to build a new tunnel here, some fifty years later, one of his most experienced contractors remarked that 'the Rock I find extremely hard, some of it in my opinion is much harder than any tunnel has ever been driven in before.') The presence of quicksand didn't help matters either.

Again, to save both effort and money the Harecastle Tunnel was built without a towpath. This meant that the boats had to be 'legged' through (a necessity duplicated in many later tunnels). This needed two people, and so bands of 'leggers' sprang up at such tunnels, touting for work. At first a board was placed across the fore-end of the boat, and the leggers lay down on each side of the boat. The two men (later in canal history it could also be women) then criss-crossed their legs and 'walked' the boat through the tunnel (the horse being led to the other side, meanwhile).

The problem was that if one legger fell off, the other did too. Separate boards called 'wings' were then developed to prevent this. But legging could take up to several hours in the longer tunnels, and – despite the droppings from the roof that might fall on top of you at any moment – it was not unknown for men, drowsy from the monotony, to fall asleep, slide into the water and drown.

As the work got going, in September 1767, a gentleman of Burslem wrote to a friend:

> *Gentlemen come to view our eighth wonder of the world, the subterranean navigation, which is cutting by the great Mr Brindley, who handles rocks as easily as you would Plumb-pies, and makes the four elements subservient to his will . . . when he speaks, all ears listen, and every mind is filled with wonder, at the things he pronounces practicable.*

As it turned out, the author had spoken too soon. So had James Brindley. At a meeting of the Trent and Mersey committee in 1767, he had personally bet a gentleman present the considerable sum of £200 that the canal would be finished within five years. He died before the time was up – but if he hadn't he would have lost his money. The canal took eleven years to finish, largely because of the difficulties in constructing the Harecastle Tunnel.

The junction between the Bridgewater extension, which Brindley was also supervising, and the Trent and Mersey at Preston Brook was finished in 1773, after running into its own engineering problems. Then calamity struck. The section of the Trent and Mersey from Middlewich to Preston Brook was built as a broad canal, so as to be able to accommodate the wide beam of the Duke of Bridgewater's boats. It was now found that the duke's boats were too broad for the new section. Each party naturally blamed the other, the engineers of the canal insisting that the duke must have increased the size of his boats since the original agreement was made.

More problems ensued when work started on the final section, from Preston Brook to the Mersey estuary at

Runcorn. A classic case of what became known as nimby-ism – and by no means the last – temporarily deadlocked the canal. The refusenik was Sir Richard Brooke, owner of the splendid Norton Priory, which he had recently, at great expense, turned into an elegant country estate, complete with palatial home. Sir Richard adamantly refused to allow the canal to cross his lands – even though this had been approved by parliament. The section thus remained unfinished for three years. The Duke of Bridgewater presented a petition to parliament, with the hope of compelling Sir Richard to accede. A fierce debate ensued in the House of Commons.

While some MPs argued that Sir Richard's estate should be sacrificed for the good of the country as a whole (this was, after all, the missing link in the whole coast-to-coast waterway), others defended the 'sacred right of property'. It was a debate that was to continue, where transport in all its forms is concerned, right down to our own day. The duke's petition was thrown out.

Finally, though, a compromise was reached between Sir Richard and the duke. The 'Battle of Norton Priory' was over, and the canal could trespass across his lands.

While there were no locks at all along the 30 or so miles of the rest of the Bridgewater Canal, in the 5 miles from Preston Brook to the estuary itself there were ten, built as two staircases of five. Wedgwood was impressed. The tidal lock, he thought, 'seems to be the work of Titans, rather than a production of our Pigmy race of beings'.

Finally, in May 1777, the Trent and Mersey was fully open to through traffic. (Sections had of course been in use as soon as they were finished.)

The Trent and Mersey Canal – the journey itself

Moorings on the Trent and Mersey canal

Robert Aickman, one of the founders of the Inland Waterways Association in 1946, wrote in *Know Your Waterways* that 'in variety the country traversed by the Trent and Mersey is equalled on no other waterway; the most beautiful districts are perhaps at the two extremities, the wooded hills at the northern end; and the spacious beauty of the Trent Valley in the neighbourhood of Weston'. In between, the section from Great Haywood to Fradley reminded him of East Anglia, while the Caldon Branch offers 'magnificent scenery: it penetrates the uncompromising but impressive uplands where Derbyshire, Staffordshire and Cheshire meet'. The canal also passes through what was the industrial powerhouse of the Potteries. Along its 93 miles are 76 locks and 4 tunnels, including the notorious one (in reality, two) at Harecastle. Of the 92 original distinctive cast iron mileposts, 59 are originals.

Travelling eastwards, our journey begins 11 yards inside the north end of the **Preston Brook Tunnel**, where the Trent and Mersey meets the Duke of Bridgewater's extension to Runcorn and the Mersey estuary. On the path above, a milepost indicates 92 miles. It's thinking of Shardlow, because there are 93.4 miles to Derwent Mouth, where the canal gives way to the River Trent. (The milestones all count down to the former.) By the tunnel – too slim for more than one narrow boat at a time – is a former dry dock, used by the steam tugs that helped haul boats through the tunnel, starting in 1865.

Immediately on leaving the tunnel at **Dutton** we're in farmland and woodland, with clumps of copses, and trees often lining the canal. From Dutton Hall the River Weaver draws ever nearer. A series of small bridges span

the canal as it approaches **Saltersford Tunnel**; 424 yards long, it was (accidentally) built with a kink in the middle, thus making it impossible to see the proverbial light at the end. 160 yards later comes the **Barnton Tunnel**, 572 yards long and kink-free. After this is a row of cottages, once homes to sailors of 'Weaver Flats'. The **Stanley Arms** is one of around sixty canalside pubs along the Trent and Mersey.

Just past it is the famous **Anderton Boat Lift**, the 'Cathedral of Canals', which lifts boats 50 feet up from the River Weaver and deposits them in the canal. Its alternative title, the 'Anderton Vertical Lift', sounds somewhat tautological, but is meant to distinguish it from the inclined plane lift. Built by Edwin Clark under the supervision of Sir Edward Leader Williams (also responsible for the Manchester Ship Canal), it was opened for business in 1875. As originally designed, boats passed into two caissons, or watertight containers, which counterbalanced each other as they went up and down. One barge or two narrow boats can fit into each compartment. Steam-driven hydraulic rams helped raise and lowered the caissons. This ingenious system never worked all that well in practice, and the machinery – perhaps because of high levels of salt in the water – became corroded. It was modified in 1908, and the two caissons are now lifted entirely by hydraulic power, independent of each other. Electricity took over in 1912. The lift was closed in 1983 but reopened in 2002, having been restored thanks to lottery funding. A **visitor centre** explores its history.

After the Anderton Lift the canal passes through **Marbury Country Park**, with its celandines, wood anemones and bluebells, and does its best to skirt around the suburbs of the salt town of Northwich to the south. Around this area centuries of salt mining have caused much subsidence, and a new section of the canal at **Marston** needed to be dug in 1958. From here to **Broken Cross** industry has left its scars. **Lion Salt Works** was closed in the 1980s, but the factory building still stands. It once ran its own fleet of boats. Nearby, indeed, is the **Salt Barge** inn. The factory of Brunner Mond, which became

part of ICI, is on the canalside after **Wincham Wharf**. 'Donkey engines' can still be seen pumping brine nearby.

After the village of **Higher Shurlach** – at the centre of the county of Cheshire – comes a series of small lakes, where, in the 1950s, bizarrely, the British Transport Commission decided to scuttle several dozen old narrow boats that were surplus to their requirements. The area is rich in birdlife. From here we float past **Whatcroft Hall**, with its copper cupola, and on through the beautiful **Dane Valley** to the **Croxton Aqueduct**, rebuilt in narrow guage after a flood in the 1930s. Nearby are more submerged narrow boats.

Middlewich is another major salt town, once home to household names like Cerebos and Saxo. Bisto was once made here. The town is by tradition a home for retired working boatmen, and the Middlewich Folk and Boat Festival takes place every June. The **Big Lock** here is so named because it's the last broad one, travelling eastwards, until we leave Burton-on-Trent. At **Wardle Lock** the very short canal of the same name joins up with what was the Ellesmere and Chester Canal, now part of the Shropshire Union.

From the industry of Middleton we now pass through a mostly rural area, running to **Wheelock**. The section of canal from here up through the Cheshire Plains to Hardings Wood, near the Harecastle Tunnel, was known to boaters as 'Heartbreak Hill', with its 26 locks in 6 miles, the canal rising 250 feet. With the exception of the pair at Pierpoint the locks are duplicated (although the side paddles that once drained into the duplicates have been decommissioned). The work was overseen by Thomas Telford.

We now pass the outskirts of the handsome town of **Sandbach.** In the market place are two famous Anglo-Saxon crosses, 16 and 11 feet high, commemorating a king of Mercia, Peada, son of Penda. He may have shared most of the letters of his name with his father, but not his religion. Peada was the first Mercian king to be baptized a Christian.

Malkins Bank was another traditional retirement spot for boaters, and there's still a yard restoring traditional working boats. From here to Kidsgrove is pleasantly rural, with pastures and woodlands surrounding the canal, and views of the **Old Man of Mow**. At **Lawton** is the fine neo-classical **All Saints Church**, rebuilt in 1798 after a fire. Telford constructed three new locks here to replace a former Brindley staircase.

At **Red Bull Locks** (not so-called because of a sponsorship deal) are old canal warehouses and an antique crane. And the border with Staffordshire. **Pool Lock Aqueduct** takes us over the Macclesfield Canal. At **Hardings Wood Junction** Heartbreak Hill comes to an end (unless you are coming in the other direction, of course). From here are views of **Mow Cop**, site of a series of famous Methodist revival prayer meetings, beginning in 1807. Nearby is **Wilbraham's Folly**, a summerhouse in the shape of a medieval castle, dating from 1754.

We are now in **Kidsgrove**, at the canal's summit, and the tell-tale orange water, coloured by ironstone, shows us that the **Harecastle Tunnel** awaits. James Brindley's tunnel, finally finished in 1777, eleven years in the making, was closed in 1914. Thomas Telford's wider version was built in 1827. Though its original towpath has collapsed, it's still navigable, although it was closed between 1973 and 1977. While both tunnels were in operation, southbound traffic used Brindley's, while boats travelling north used Telford's. Neither tunnel could take more than one narrow boat at a time.

One famous employee of the Telford tunnel is 'Kit Crewbucket', a corruption of the 'Kidsgrove Boggart', or phantom. The ghost of a decapitated woman is also said to haunt the tunnel. (There is no record of such a murder taking place.) For good measure the spectre of a white horse is also supposed to get in on the act. From the horse path on top of the tunnel – 'Horseboat Lane' – Jodrell Bank can be seen on a clear day.

After leaving the tunnel we pass through **Tunstall**, one of the six Potteries towns that make up the modern conurbation of Stoke-on-Trent. (Confusingly, local author Arnold Bennett wrote a famous novel called *Anna of the Five Towns*. Perhaps

arithmetic was not his strong point.) Tunstall was once an important iron centre, but now only the ravages caused by the former industry remain.

Westport Lake, to the south of the canal, was opened in 1971 and is a popular resort for the local inhabitants. It was once the home of Port Vale football club – until it sank, creating the lake. It later became a rubbish dump. Today it's an attractive setting, the biggest stretch of water in the city, and a haven for wildlife.

The once-busy Canal Street, on the west bank in **Longport**, is now mostly derelict. But opposite, near Longport Bridge, stands one of the distinctive old 'bottle kilns' that once dominated the district. This one is the 'Price Kensington Tea Pot Factory'. There are a few more along this industrial section.

As we approach Wedgwood country we pass through the now abandoned site of what was until 1978 the massive **Shelton Bar** steelworks, which once had 10,000 employees. Robert Aickman, writing in 1955, said of it, 'The passage during working hours through the Shelton Iron Works is something to remember for a lifetime.' The **Festival Park** on the north bank was formed from what was the Shelton Works, with part of the former Wedgwood factory at **Etruria** added. Only a solitary roundhouse, once one of two guarding the entrance, remains. **Etruria Junction**, from which the **Caldon Canal** heads north, commemorates this once-mighty factory. (In the 1940s, largely because of subsidence, it moved east to Barlaston, where we will encounter it in its modern incarnation.) The **Etruria Industrial Museum** is on the Caldon Canal, as is a statue built in honour of James Brindley – along with Josiah Wedgwood, the presiding genius of the Trent and Mersey.

At **Trentham** are the recently restored Italianate gardens, designed by Charles Barry (one of the architects of the Houses of Parliament), that once belonged to Trentham Hall, the seat of Earl Gower, another crucial figure in the canal's development. (Charles Bridgeman and Capability Brown had also been involved in the earlier landscaping.) It was demolished in 1912. The gardens were gifted

to the city of Stoke. Recently developed, they contain a monkey forest, a 60-acre site that's now – somewhat unexpectedly – home to some 140 Barbary macaques.

Nearby at **Barlaston** is the present-day Wedgwood factory, opened in 1940. It contains a **visitor centre** showcasing the history of this fascinating and hugely important historic firm.

After **Meaford Locks** we leave the Stoke conurbation to travel through some open countryside, towards **Stone**. Roughly at the halfway point of the journey, Stone was once the HQ of the Trent and Mersey Canal Company. The town has several fine canalside buildings, including the **Joules Brewery** and **Star Inn**.

From Stone we pass through a celebrated stretch of the canal, with gently rising farmland and scenery of great beauty. The **Trent** runs nearly parallel, and there are several attractive bridges as we make our way through the quiet village of **Burston** and reach the picturesque one of **Sandon**. The **Dog and Doublet** is a famous canal pub. At **Sandon Hall**, set in pleasure grounds high above the canal, is a monument to Spencer Perceval, the prime minister who was assassinated in 1812. Within the woodlands of the park there's another, dedicated to William Pitt the Younger. From **All Saints Church** there are good views of the Trent Valley.

After **Weston** we pass **Shirleywich**, another salt centre, and, to the south, the nature reserve of **Pasturefields Saltmarsh**. The countryside from here to the outskirts of Rugeley provides superb scenery. The Jacobean **Ingestre Hall** is today a residential arts centre, its gardens planned by Capability Brown. At **Haywood Junction** we meet the Staffs and Worcs Canal, overlooked by a Grade II listed roving bridge.

We now pass by the imposing stately home of **Shugborough Hall**, owned by the National Trust. It was designed by Samuel Wyatt in the early eighteenth century for the family that would become the Earls of Lichfield. In the park are several classical follies, in the 'Greek Revival' style, built by James 'Athenian' Stuart. At the River Trent, now right next to the canal, is the beautiful **Essex Bridge** (named after the Earls of Essex, who once owned land here). An ancient monument, the longest packhorse bridge in the country (fourteen of the original forty spans survive), it gave the earls access to the hunting grounds of Cannock Chase to the south, now designated an Area of Outstanding Natural Beauty.

At **Colwich** is **St Mary's Abbey**, established by Benedictine nuns in 1836. J. R. R. Tolkien lived nearby while recuperating from trench fever during the First World War, and the countryside here is said to have inspired his early descriptions of 'Middle-earth'. The present **Wolseley Bridge** was designed by John Rennie. The **Wolseley Arms** was a regular meeting place for Wedgwood, Brindley, the Gilberts and other movers and shakers of the canal's earliest days. Nearby is **Swan Lake**, a pretty nature reserve.

From here the canal becomes far less rural. From **Brindley Bank** we come to **Rugeley**, a former mining town. Near Armitage Tunnel is **Spode House**, the former property of the important potter Josiah

Spode, followed by the home of Armitage Shanks. After **Handsacre** we are back in an area of pleasant scenery, with oak, birch and alders lining the canal. From **Wood End Lock**, the most southerly point of the Trent and Mersey, Lichfield Cathedral can be seen. After passing by Fradley Wood we come to **Fradley Junction**, where we meet the Coventry Canal. Here are several attractive converted warehouses and a famous, 200-year-old canal pub, the **Swan**, or 'Mucky Duck'. Further along, past more gentle farmland, is **Alrewas**, an attractive village whose name is a corruption of 'Alder Wash'. Here there is a weir besides the River Trent, which briefly joins the canal. At Alrewas we enter the planned 200-acre **National Forest** – a project to meld ancient woodland with new plantings and recreate large areas of Midlands forest long ago lost. (The A38 now makes its presence felt too, dogging the canal right up to Burton-on-Trent.)

At **Wychnor** are a thirteenth-century church and half-timbered cottages. A flitch of bacon was, in former times, on offer to any man who could swear that during his first year of marriage he had never thought of trading in his wife for another model. Sadly, the prize doesn't seem to have ever been awarded.

At the wharf at **Barton Turn** several old canal buildings have been put to new use, while there are pretty cottages at **Tatenhill Lock**. Next is the village of **Branston**, once home to the celebrated pickle. The aroma of hops now announces that we are entering the centre of British brewing, **Burton-on-Trent**. As is so often the case in the history of alcoholic beverages, monks were the main driving force here, those at **Burton Abbey** discovering that the water that flows through the gypsum rocks nearby (not the water of the Trent, as is sometimes thought) made an excellent ingredient for beer. **Sinai Park**, near the canal at Shobnall, was their summer home. In its heyday there were thirty-one breweries in Burton. One that still survives, **Marstons**, has a visitor centre on the canal at the Albion Brewery. It's the only British brewery still using the traditional methods.

At **Shobnall Basin** used to be found the **Bond End Canal**, which connected to the River

Trent, where most of the breweries were sited. From here beer was transported to Hull and Liverpool. The famous India Pale Ale, initially only made for export, was a favourite tipple in the days of the Raj.

Dallow Lock is the last narrow lock on the canal heading eastwards. Near **Horninglow Wharf** is the **National Brewery Centre**, which incorporates the Bass Museum. William Bass founded his famous brewery in 1777, having sold his carrying business to Pickfords. Leaving Burton-on-Trent, the canal passes over the **Dove Aqueduct**, constructed by James Brindley. At 79 yards long, with 9 arches, it's the longest on the canal. The River Dove marks the border with Derbyshire. After the town of **Willington** comes **Findern**, with its pretty village green, surrounded by thatched cottages.

Stenson is the first broad lock we encounter, travelling eastwards, 14 feet wide, and an attractive one. From here to Weston is fine scenery, with glimpses of the Trent Valley. At **Swarkestone Lock** the Derby Canal makes its way north, overlooked by an old toll-house, now

a boat club. At **Lowes Bridge** on the River Trent in Swarkestone, the longest stone bridge in England, Bonnie Prince Charlie was reluctantly persuaded to abandon his march on London and turn back to Scotland, in 1745. This was the southernmost point of the Jacobite army's progress.

In the pretty village of **Aston-on-Trent**, near **Aston Lock**, is **All Saints**, first established in the seventh century and much rebuilt, noted for its Victorian stained glass. After passing through the lock we enter the attractive inland port of **Shardlow**. Old canal buildings still stand, including a lock-keeper's cottage and the elegant **Clock Warehouse**, built in 1780 and now a pub, having formerly been a corn mill. The 'boat hole' under the building, where boats would load and unload, still survives. The **Malt Shovel** inn, on the **wharf**, dates from 1799. At the wharves canal boats would trans-ship their cargoes on to river barges. In 1839 a petition signed by 140 boatmen called for a proscription on Sunday working. The town wasn't known for excessive sobriety, however. Reputedly, it once had the largest number of pubs per head of population anywhere in Britain. Just to the east of the town, the canal gives way to the Trent at **Derwent Mouth Lock**, where the river is also joined by the Derwent, coming down from Derby. From Derwent Mouth the Trent winds north through Newark and Gainsborough to Trent Falls, where it meets the Yorkshire Ouse and drains into the Humber estuary.

Not only were canals at the heart of the industrial revolution. Their owners and directors were at the heart of the capitalist system. The Duke of Bridgewater, Earl Gower, the Gilbert brothers, Brindley, his brother John and brother-in-law Hugh Henshall, the clerk of works of the Trent and Mersey, owned between them a complex group of interlocking companies, coal mines, iron works, mills and farms.

Hugh Henshall was also the nominal proprietor of the Trent and Mersey's own 'carrying company', which owned and managed a fleet of boats and charged clients for transportation of their goods. This was rare in the early days of canals. Although the Duke of Bridgewater owned his own boats (and indeed every part of his canal, lock, stock and barrel), more usually the canal company simply owned the 'track' and allowed other business interests to operate boats on the canals, sitting back and pocketing the tolls without the need for getting involved in what became an increasingly competitive and hard-to-manage field. The canal companies were responsible for maintaining the canals (in theory, at least).

There was, in fact, a devious reason why the Trent and Mersey favoured keeping its own carrying company at arm's length. The Acts of Parliament authorizing canals set the level of tolls and insisted that they could not vary along the whole length of the canal. A carrying company, on the other hand, could alter its fees depending on supply and demand. Thus the Trent and Mersey could milk the more profitable sections.

The canal, as it was designed to do, took the pottery trade out of the hands of the Weaver Navigation but failed to capture its salt trade (largely because of the problems caused by the final section of the canal not being broad enough, and the duke's failure to build an aqueduct over to the Lancashire bank). But a connection, with trans-shipment dock, was built at Anderton in 1793 (later to be replaced by the gigantic Anderton Boat Lift).

Coal became an ever more important commodity as industry moved over from water to steam power. Many other factory owners followed Wedgwood's lead and built new premises by the side of the canal, where previously they had been sited next to fast-flowing streams.

Industrialists built their own wharves. The fact that they could now count on regular supplies of raw materials had a great economic benefit. They didn't, as in days of old, when the arrival of supplies was harder to predict, need to order such huge quantities, thus tying up both capital and factory space.

The Grand Cross Takes Shape

The Trent and Mersey had done what it set out to do and brought huge benefits to trade. But it was not the first spar of the Grand Cross to be completed.

On the same day the Trent and Mersey received its

Authorizing Act, so did another canal company, which also employed James Brindley as chief engineer. The Staffordshire and Worcestershire Canal joined the Trent and Mersey at a junction at Great Haywood, east of Stafford. Though they were separate enterprises, the two companies enjoyed good relations. They were, after all, both part of the same master plan. (Future canal companies were to be considerably less friendly towards one another.)

This happy state of amity wasn't duplicated on the Severn. Brindley's plan had been to link his canal to the river at Bewdley. From there it was navigable down to the Bristol Channel. The Severn was, however, a hazardous river to negotiate. Consequently, the boatmen of Bewdley made a good living piloting boats through the notoriously tricky tides. So they turned their noses up at giving house room to what they called Brindley's 'stinking ditch'. Instead, he chose a point four miles further downstream. An entirely new town, Stourport, was created, and its fortunes thrived while those of the intrepid navigators of Bewdley declined.

The Staffordshire and Worcestershire was, because it linked with the Trent and Mersey, a narrow canal. Indeed, Compton Lock was the first narrow lock that Brindley built, with a width of seven feet. (The 'pounds' between locks were far wider, of course, allowing boats to pass.) From its completion in 1772 the Staffs and Worcs was highly successful. Originally referred to as the 'Wolverhampton Canal', it took coal and iron from the Black Country down to Gloucester and Bristol, where it could now compete against that of Shropshire. Via the Trent and Mersey, pottery and other goods also came down to the same cities.

Little wonder that at the new transport hub of Stourport, as a historian of the next century put it, 'Houses, warehouses and inns sprang up as if by magic, the magic which wealth-creating industry usually gives, and iron foundries, vinegar works, tan-yards, spinning mills, carpet manufactories, and boat-building establishments were added.'

Such was the astounding success of these early canals that in the nineteenth century authors like Samuel Smiles and Thomas Carlyle tended to hero worship James Brindley, the 'workman genius'. He was a classic case, of course, of 'Self Help', raising himself up by his boot-straps by dint of his innate genius and Stakhanovite work ethic. But this is a little unfair. The reason his canals were, for the most part, successfully brought into being is down, not just to Brindley, but also to the talented engineers whom he trained and wisely surrounded himself with.

This close-knit crew included Hugh Henshall, Brindley's brother-in-law. Not only did he run the carrying company for the Trent and Mersey. Long before this he had helped Brindley survey, and drew the map for Earl Gower's original Trent canal. He took over the engineering of the western end of the canal on Brindley's death, building a series of tunnels through the marshy ground above Middle-wich. Also closely involved was another of Brindley's brothers-in-law, Samuel Simcock, and a further brother-in-law of Henshall, Josiah Clowes. Robert Whitworth, who married Brindley's widow, completes the set. All of these men were in day-to-day control of much of the canal network, under Brindley's supervision. All were involved in finishing off Brindley's canals after he died.

It was this band of brothers-in-law, Brindley's so-called

'School of Engineers', that worked out how to build a canal and set the pattern for those that followed.

Building the Canals

The basic techniques of surveying in the eighteenth century are still familiar today (except that Brindley and his surveyors were working before Ordnance Survey maps had been published: the first was seen in 1801). Surveyors and their assistants used spirit levels and, later, theodolites. 'Triangulation' was employed to determine the height of any object that needed to be cut through. They calculated distances using chains (which became standardized at 22 yards). Rods would be set to show the depth of the canal to be cut.

Once work was underway, the early canals were dug, by and large, with pick and shovel, blood, sweat and tears (although some mechanization was used later). This backbreaking task was done by unskilled labourers. Earth was dragged up from the cutting, sometimes with the help of horses operating pulleys, quite often by a man pushing a wheelbarrow. On deeper cuts wooden boards were laid down, and the slippery surface often caused both barrow and workman to career straight back into the cut.

The puddling, while the principle was simple, was less easy to apply. Labourers in heavy boots had to trudge it into the floor of the canal and press it into the sides.

Brindley constructed the canal length by length and built a temporary 'stop-lock' or barrage so that sufficient water could then be added for boats to take away the surplus earth, as well as bringing materials for construction. Meanwhile the engineers would use boats as floating bureaus so that they could, literally, keep up with the progress of the work.

The hiring and supervision of the labourers was assigned to contractors. The work was subcontracted out in parcels, in job lots of so many locks, so many bridges or tunnels and such a length of canal. The records of the canal companies, starting with the Trent and Mersey, reveal almost daily battles between the clerk of works, or company engineers, with the contractors, or their subcontractors. Many were the disputes over workmanship and money that then had to be adjudicated.

The building of tunnels was a potentially dangerous endeavour, and deaths and accidents did occur.

Locks posed their own problems and had, of course, to be built to a high standard to conserve the precious supplies of water. 'Pound-locks' – a pound is the stretch of water between two locks – had already been successfully developed by the engineers of river navigations (although Brindley supposedly wanted to design his own and was rumoured to have built a prototype in his back garden).

The men who cut the locks were better-paid than the ordinary labourers, as the work required greater accuracy. Having been 'puddled', the locks were then faced in masonry, while the gates were made of good-quality wood, usually oak. And so a small army of masons and carpenters was involved in the construction of the canals. (Reputedly, one of the most soul-destroying jobs of all was 'water-scooping' – getting rid of the water from the bottom of the lock so it could be lined.)

Locks were often built in pairs, the bottom gates of the

upper lock also acting as the top gates of the lower. This is known as a 'double lock' or 'riser' (although the term 'double' is sometimes, misleadingly, applied to two locks side by side, more accurately known as a 'duplicate' lock – these were a later development). A 'staircase' usually refers to more than two locks that joined together. (Curiously, on the Staffs and Worcs Canal, Brindley seems not to have grasped this principle, and on the set of three locks at the Bratch he built extremely short pounds between them. The locks he later built at Botterham, however, are double locks, joined together without an intervening pound.) A 'flight' consists of two or more locks in close proximity to each other but separated by pounds.

Lock gates are 'mitred' – an invention often credited to Leonardo da Vinci. They close to form a wedge, the apex of which always faces the summit of the canal. Thus the gates offer more resistance to the water above. Guillotine locks are also found, but rarely. One was built at Kings Norton, on the Stratford-upon-Avon Canal, and there were nine small guillotine locks on the Shropshire Union.

Locks were for the most part designed to accommodate, as tightly as possible, only one of the principal craft the canal was made to be used by, so as to maximize water. But there were some eccentric exceptions, such as lozenge-shaped locks at Wyre, on the Lower Avon Navigation, and Cleeve, on the Upper Avon. Cherry Ground Lock, on the River Larke, was made in the form of a crescent moon, while, even more bizarrely, Luddington Upper Lock on the Upper Avon was circular.

Towpaths were nearly always built on just one side of the canal. This meant that when two boats passed, one of them had to drop the tow line into the water so the other boat could float over it. Passenger steamers always had the right of way here – or were meant to, at least.

When a towpath crossed over to the other side (perhaps because of a dispute with a landowner) a 'roving bridge' was built to allow the horse to get across. (A later refinement was to build a bridge with a slit down the middle, so the tow line could be passed through without having to be removed from the boat.)

Tunnels, of course, meant that horse, and owner, had to find their own way to the other end.

Canals had, so far, carried all before them and were an undoubted boon to the communities they served. But, as so often in the annals of history, success brought with it less sunny aspects of the human character. All was not sweetness and light on the canals for long. As their great chronicler, L. T. C. Rolt, put it, in *Navigable Waterways*,

> *as soon as the profitability of canals was fairly proven, middle England became a Tom Tiddler's ground of rival canal companies, each jealous of their territory and intent to parry any rivals who threatened to siphon off any part of their traffic, the trade in coal being the most zealously fought for.*

Nowhere was this to be proved more true than in the tangled, fractious and tortuous history of the Birmingham Canal Navigations.

The Birmingham Canal Navigations

Introduction by John Sergeant

Birmingham owes more to its canals than any of the other large towns in the country. Its network of waterways enabled it to become not only England's second city, but one of the greatest manufacturing centres in the world; and this was despite its almost total lack of raw materials and its distance from the sea. Birmingham used to be described as the town of a thousand trades, and their success was only possible because of the rapid growth of canals at the end of the eighteenth century.

Usually for this television series I concentrated on a single canal route, but that would have been too restrictive. I wanted to look at the hub of the canal system, with Birmingham's many canals radiating out from its centre. At times this could be confusing. Signposts are provided at the various junctions, but it is not easy to get to grips with the whole network.

At one point I was under the M6 motorway at a meeting place of three different canals going off in different directions. There was a constant rumble of road traffic overhead as we made our way under bridges daubed with graffiti. It was hardly a tourist destination, but it was a striking example of how transport systems of all sorts can get into a bit of a tangle. In our narrow boat we were going right under Spaghetti Junction. In some places modern roads and old canals are not that different.

It was, of course, the industrial revolution that put Birmingham on the map. Two of the great founding fathers were James Watt, whom some regard as the finest engineer Britain has ever produced, and his astonishingly successful business partner, Matthew Boulton. They are commemorated in the city centre by a statue appropriately bedecked in bright gold. Every time you check on the power of an electric light you are paying homage to Mr Watt.

In a suburb we went to see the remains of their Soho Foundry, which was the only factory producing the great steam engines that made their fame and fortune. Almost all the original buildings have been replaced, but a small opening on the canal shows where the barges would set off with their products to conquer the world. The engineering company Avery took over the plant, and we were shown round a small museum, with its bust of James Watt and a brilliant collection of old weighing machines. Many of them carry the proud label 'Made in Birmingham'.

The city's engineering prowess became legendary in many fields, especially in military equipment. The giant factories, which produced the rifles and artillery to arm the Empire, are long gone, but at the small plant run by Westley Richards & Co., founded in 1812, precision-made firearms are still being built by hand. The market for these intricate, beautiful objects is largely confined to the very rich; and often they go straight into collections abroad. A presentation gun, with a diamond-encrusted stock, sells for a cool £250,000.

The centre of Birmingham has been transformed in recent years. The Gas Street Basin used to attract visitors to the hub of the canal network because it was one of the first places to have gas street lamps. Now it has become fashionable with its string of restaurants and bustling nightlife. Some of the modern structures, with their acres of glass and uniform red bricks, are not to my taste, but the renovations to the old buildings are largely successful, and at least pride in the city's heritage is on display. After all, they do have more canals than Venice.

The Birmingham waterways fall easily into the best of Britain category. But for the boating enthusiast they do require planning and a degree of fortitude. There are long stretches where you feel more like an explorer than a visitor. Dotted along the complex network are some real gems . . . if only you can find them.

The coming of canals would make Birmingham the most important crucible of the first blast of the white heat of technology. The 'Birmingham Canal' was the first thread in what would become a whole skein, 160 miles of canals at the network's peak, that would shift the centre of gravity of the English canal system from the north-west to the Midlands. Canals would play a vital part in making Birmingham the manufacturing powerhouse of Britain, the number one workshop of the western world, and one of the great success stories of the industrial revolution.

A century or more ago, it was generally agreed amongst historians that such a revolution had started around 1760 and really fired up around 1780. There is no doubt that in the 1780s the volume of manufacture, trade, agriculture and population in Britain all increased (the costly American War of Independence having finished in 1783). Why these increases should have taken place at the same time, rather than one triggering the others, is a matter of much debate amongst historians. The increase in trade with North America and the West Indies, particularly after the Seven Years War (1756–63), the economic and financial stimulus of almost constant military activity – Britain was at war with France for a third of the eighteenth century – and the wealth secured by the disgusting trade in slaves have all been suggested as enabling factors. But the Brindley canals that were finished by the 1780s must surely be counted as another.

While it's agreed that the 1780s saw a shifting up through the gears, historians over the last hundred years have pushed the starting date of the industrial revolution further and further back in history, with a broad consensus locating its origins in Elizabethan times (when coal first began to be used instead of wood), while some antecedents reach back to the late medieval period. Certainly there was no starting pistol fired. It's come to be seen as an industrial evolution rather than revolution.

The history of Birmingham's population growth charts the progress of this sea change. In 1538 it was a town of 1,300 people. A hundred years later it had more than tripled in size. In 1700 the figure stood at around 15,000. By 1778 this had nearly tripled again. In 1801 (the year of the first proper census) the population was 73,000, which was to double again by 1831. A century later Birmingham (a city since 1889) was home to over a million people.

It's not hard to see what caused this boom in numbers. Birmingham became known as the 'City of a Thousand Trades'. And that might have been an underestimate. Birmingham was blessed – or cursed, from a clean air point of view – with large quantities of coal, limestone, ironstone and clay right on its doorstep. There were extensive quarries at Dudley and further west, while the rest of the Black Country, around Wednesbury, Wolverhampton and Walsall, was a rich source of coal. The Warwickshire coalfield was east of Birmingham: at Tamworth, Coventry and Nuneaton. To the north was the deep coal of Cannock

Chase. Ironstone was also to be found in abundance in the hinterland of the town.

Coal, iron and limestone (used in iron foundries, construction and agriculture) were three vital ingredients of the industrial revolution. Water – for transport and initially for power – was the fourth.

The Birmingham Canal in the 1920s

By the time the canal was proposed, in 1767, Birmingham was already a thriving industrial town, known particularly for its metalwork. The fact that it had 'just growed' like Topsy meant there was no guild system to regulate – and restrict – manufacture and trade. Instead, there was a free-for-all, and the cut-throat world of market capitalism held sway. (Banking facilities were provided by the Quaker families of Yorkshire and East Anglia.)

However, like Wedgwood and his fellow potters, the metal-makers of Birmingham were hamstrung by lack of transport. The town is on a plateau, 350 feet above sea level. Roads were poor, and the rivers were unnavigable, either too shallow or too fast-flowing. What's more, the millers had been at work and had siphoned them off to drive their waterwheels. So the making of river 'navigations' was a non-starter. In the 1760s, eyes turned to the construction of a canal.

It wasn't only manufacturers who were in favour. A symptom and a cause of Birmingham's rise to prominence as an industrial and commercial centre was the combination of entrepreneurs and *savants* who worked together, actively seeking opportunities for the town to thrive. This was, after all, the Age of the Enlightenment, with its interest, not just in philosophy and politics, but in science, and the possibilities it offered for new technology, and thus increased trade.

The 'Lunar Society', founded in 1765, was an informal group of Birmingham scientists, inventors and industrialists who met every full moon (making it easier and safer to get home in the days before street-lighting – the 'Lunaticks', as they styled themselves, were highly practical men). The society became the intellectual engine room of the movement known to history as the 'Midlands Enlightenment'. Members would include Erasmus Darwin, Joseph Priestley, who discovered oxygen, the manufacturer Matthew Boulton, one of the key figures of the industrial revolution, and, later on, his business partner, James Watt, ditto.

Josiah Wedgwood, while he may not have been a formal member, was a regular guest. This powerhouse of scientific and business interests sponsored research, experiment and invention. And they were all very much in favour of a Birmingham canal. (Canals went on to play a small part in pure science. Josiah Wedgwood sent Erasmus Darwin a number of fossils that had been found while the Harecastle Tunnel was being dug. They would help Darwin to arrive at an early version of his grandson's theory of natural selection. The different strata that were revealed in deep cuttings would also aid the new discipline of geology.)

The Trent and Mersey Company had proposed building a Birmingham branch, but the offer wasn't taken up. Now, seeing the success of the Bridgewater, Trent and Mersey, and Staffordshire and Worcestershire Canals, the business leaders of Birmingham decided to put their own project in train.

Erasmus Darwin
(1730–1802), inventor

In June 1767, a canal for Birmingham was first discussed, at a public meeting in the Swan Inn, West Bromwich (inns being the conference centres of their day). A subscription book was opened, and contributions were sought to pay for the costs of presenting a petition to parliament. Influential Birmingham worthies stumped up, including Matthew Boulton, the chemical manufacturer Samuel Garbett and the gunmaker Samuel Galton. John 'Iron-Mad' Wilkinson, one of whose foundries was at Bradley, north-west of Birmingham, was also a member of the original committee. They quickly raised £50,000, and James Brindley was engaged to conduct a survey.

Opposition from local coal-owners immediately surfaced. They feared that cheaper coal would be brought in by water and undercut their own, which arrived by the expensive means of horse-drawn transport, or by primitive tramroads. They were right. As the Bill presented later the same year put it, 'the Primary and Principal Object of this Undertaking was and is to obtain a Navigation from the Collieries to this town'.

An Act was passed in 1768, and a route was agreed from Newhall, in the centre of Birmingham, north-west to Smethwick, on through the collieries at Oldbury, Tipton, Bilston and Wolverhampton, and thence to Aldersley, where the Birmingham Canal would join the Staffordshire and Worcestershire Canal. Here boats could head for the Mersey in one direction, the Severn in the other. Branches would be built to Wednesbury, Ocker Hill and other collieries in the Black Country area.

Brindley was the supervising engineer, with the work on the ground done by two of the leading members of his 'School of Engineers', Robert Whitworth and Samuel Simcock (who was married to Brindley's sister).

Even by the time the first section was completed, the ten miles from Wednesbury to Paradise Row (as Paradise Street was then known), the enormous value of the canal was immediately apparent. A local poet (also a member of the canal committee) summed up the reason for this mood of unbounded glee:

> *Then revel in gladness, let harmony flow*
> *From the district of Bordesley to Paradise Row,*
> *For true-feeling joy in each breast must be wrought*
> *When coals under fivepence per hundred are bought.*

Coal had been selling at more than three times this amount. By 1770 the town was awash with coal, and steerers, carters, carts, boats and horses were sought far and wide. The company's boats now worked through the night, as did the lock-keepers, toll-collectors and wharfingers. It was a case of money no object to get more canal built as soon as possible. Brindley was instructed to drive ahead at full speed.

The Birmingham Canal Company had begun with pious assurances to the town that it would not profiteer from the trade in coal. Indeed it made a point of selling small amounts of it to the poor at the wholesale price. But then the philanthropic impulse seemed to wither on the vine, and some strange and disturbing goings-on were noted by the company's growing band of critics. The line agreed by parliament now started to take on a mind of its own, wilfully missing out some of the collieries that it had been agreed it would pass by, instead seeking out others that had not been on the original plans. So meandering was the canal, even by Brindley's standards, that after he died and Samuel Simcock took over local humorists accused him of trying to carve his initials through the town. The colliers who found themselves left high and dry thanks to the canal's mystery tour were not best pleased.

Six more miles of canal had been built in addition to the sixteen miles that parliament had agreed to. Brindley himself was taxed with deliberately planning this serpentine route so as to increase the amount of tolls, at the expense of rapidity of transport. He argued, with some reason, that the canal thus served more customers. But why, the critics wanted to know, were some customers serviced while others weren't? What mysterious factor could have determined the choice as to which collieries the canal was willing to honour with its presence?

A tide of criticism fell on the company. It was accused – entirely justly – of favouring some colliery owners over others and straying from the agreed parliamentary line; of trespassing on premises it had no permission to enter; and of building its canal on land it had no authorization to pass through. It was charged with trying to corner the market, especially in coal, by boosting its own boats, wharves, warehouses and carts and muscling out independents, as well as refusing collieries and businesses free access to the canal and the right to dig basins and branches.

In 1769 Samuel Garbett, one of the main sponsors of the canal, resigned from the committee and became its most fervent enemy, promoting meetings to broadcast its iniquities and publishing letters and pamphlets calling for its wholesale reform. His view was that the canal was meant to be, and should continue to be, for the good of the trade of the whole town, not a vehicle for filling the shareholders' already ample pockets. In 1770 he became one of the backers of a rival canal that aimed to run from the Walsall collieries – not then served by canal – to Lichfield and then up to Fradley, on the Trent and Mersey Canal. Opposition from landowners (as well as the Birmingham Canal committee) put a stop to the idea.

The same year, the company's growing notoriety had reached the ears of George III himself. A London merchant told the king that the shareholders were siphoning off profits which should have been spent on improvements to the canal, to the benefit of the populace at large.

Such was the strength of criticism, the company realized it would have to be seen to be doing something. It promised to allow private branches to be made so long as they did no damage. It promised to suspend night-working and make some improvements to the canal. Two new reservoirs were built, and the water supply was increased by 'back-pumping' – taking water from downstream of a lock, or flight of locks, and pumping it back, through a culvert, to the upper pound. But the criticism continued.

The Staffordshire and Worcestershire Canal Company was also growing restless. The proprietors had managed to get a clause inserted into the Act that allowed their company to build the north-western section of the Birmingham Canal, down from Aldersley, if the Birmingham company failed to complete it – at the latter's expense. The Birmingham outfit was dragging its feet. Great pressure was put on the committee, and the Birmingham Main Line was completed in 1772.

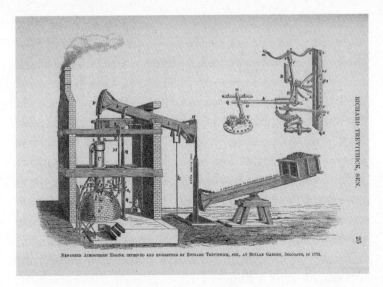

NEWCOMEN ATMOSPHERIC ENGINE, IMPROVED AND RE-ERECTED BY RICHARD TREVITHICK, SEN., AT BUILAN GARDEN, DOLCOATH, IN 1775.

A 1775 Newcomen engine, adapted by Trevithick

The canal, not including branches, was 22½ miles long. It had narrow locks of 7 feet wide and was around 28 feet wide in the pounds. (By 1795 they had been widened to around 40 feet.) The canal was a financial success. Toll receipts and dividends rocketed, tripling in just twenty years.

Steam Power

The reason Birmingham industries went into overdrive from the 1780s was due not only to the Viagra effect provided by the canals, but also to one of the most crucial inventions of the industrial revolution. Its genesis may not have been in Birmingham, but Birmingham was where it took its first confident steps into the world.

Matthew Boulton stands between Josiah Wedgwood and Henry Ford as one of the most important pioneers of the manufacturing process, instituting assembly lines and fine-tuning every aspect of production. Like Wedgwood, he was quick to realize the enormous potential of canals, and the Birmingham Canal served his own Soho Manufactory, built in 1766. It had its own Soho Wharf. Here Boulton made metal 'toys' – small objects such as buckles, hinges, buttons and hooks. He also made decorative items in japannedware (objects treated with a black lacquer), silver and ormolu (a kind of gilded bronze). In 1788 he began the Soho Mint, making coins (then in short supply).

However, Matthew Boulton's name will forever be associated with a different kind of industrial enterprise. One that involved his famous partner, James Watt.

Watt took out his first patent for a steam engine in 1769. He didn't invent the device. The notion of steam power had been around for a couple of millennia. The first working steam-powered engine had been patented and built by Thomas Savery in 1698 – *The Miner's Friend or an engine to raise water by fire*, as a book he wrote about it was titled. This was a steam-powered vacuum pump, with no moving parts (so not, strictly speaking, an engine at all). It operated by sucking up water from the surface then using steam power to raise it further. But it wasn't strong enough to be able to pump out the water to a sufficient height. This was achieved by the invention of the first steam engine to use pistons. It was patented by a Cornishman, Thomas Newcomen, in 1712, and could raise water over 150 feet. The engines were first put to use to pump water from tin and coal mines.

The drawback of the Newcomen engine was that it was 'atmospheric', involving the constant heating and then cooling of the cylinder containing the piston. Which was highly wasteful of fuel. This wasn't a particular issue for coal-owners, who had tons of the stuff to spare, but it limited the economic effectiveness of the steam engine elsewhere. The solution to this drawback was supplied by Britain's greatest ever inventor, James Watt.

Matthew Boulton
(1728–1809)

James Watt (1736–1819),
engineer and inventor

Watt was born in Greenock, on the Firth of Clyde, in 1736. The grandson of a maths teacher, son of a shipwright and carpenter, Watt was educated at the local grammar school and then trained as a mathematical instrument maker. At the age of twenty-one he set up a business making scientific instruments for Glasgow University, where he was able to study chemistry in his spare time. In 1763 he was asked to repair a model of a Newcomen engine owned by the university. Naturally, he began to ask himself how the efficiency of the machine could be improved.

The answer came to him one afternoon in 1765. Not through watching a kettle boil, as legend used to have it, but while on a stroll across Glasgow Green. A kettle may have given him a clue, though, because the main element in his new idea was a condenser, a separate container which would allow the steam to cool, and thus condense, without the need to let the cylinder itself cool down and then have to reheat it.

Watt had built a prototype in a matter of weeks, but several frustrating years followed before the new steam engine was commercially viable. He went into partnership with John Roebuck, an ironmaster who co-founded the Carron Iron Works at Falkirk, in 1760. (Samuel Garbett was another founder.) But the Watt engines couldn't be made to work. In an age before machine tools could produce accurately bored machine parts, the primitive methods then available led to defective workmanship, which caused leakages, rendering Watt's improved engine next to useless.

William Murdock (1754–1839), engineer and inventors

Roebuck went bankrupt (although the Carron foundry survived without him). Watt, now in debt, had to face the ignominy of becoming a canal surveyor. It looked like the revamped steam engine was dead in the water.

The man who saved it was Matthew Boulton. He understood the revolutionary potential of steam power, and bravely – given Roebuck's earlier experiences – backed Watt, despite the fact that the technology did not yet exist to make his engine with the precision needed. The vital breakthrough was made by none other than Iron-Mad Wilkinson.

Around 1774 Wilkinson developed a machine that could bore cylinders to a greater degree of accuracy than ever before, thus getting around the problems of the imprecise working parts which had held Watt back for nearly a decade. Now the new machine was up and running.

In 1776 Boulton and Watt built their first three steam engines. Thanks to its separate condenser the device used 75 per cent less fuel than did Newcomen's. The firm charged a fee based on the amount of coal saved, and thus the first area of take-up was in the tin mines of Cornwall, where the Boulton and Watt engines started to replace the old Newcomens. (Even the profligate coal-mine owners later began to appreciate their greater efficiency.) In 1782, Watt's introduction of a double-acting engine, in which steam was applied to each end of the piston, made them even more powerful.

The first Boulton and Watt products were beam engines, in which pistons drove the beams up and down in alternation. This was ideal for pumping, and indeed many Boulton and Watt engines were used by the canals for this purpose. The next major development in steam power was to find a way for the beam engine to turn a wheel and so drive a machine.

The problem for Watt was that the obvious method of using a crank had already been patented. And so he was forced to come up with a different means of making a rotary device. The solution was invented not by Watt (though he claimed the patent, in 1781) but by his talented foreman, another Scot, William Murdock (for a long time one of the forgotten heroes of the industrial revolution). The system became known as 'sun and planet gear' – a driveshaft turned by two rotating and interlocking cogs. (Once the rival patent ran out, the far simpler crank was utilized on subsequent rotary steam engines. Watt's further development of the 'governor' kept speed constant.)

'The introduction of steam-power,' wrote the historian C. P. Hill, 'was the greatest revolution in economic history since the discovery of agriculture in prehistoric times.' Matthew Boulton was well aware of it. When Erasmus Darwin came to visit his manufactory at Soho, Boulton told him: 'I sell here, sir, what all the world desires to have. Power.' (The famous Boulton and Watt Soho Foundry was constructed in 1795.)

The firm, over the next quarter century, supplied around 500 steam engines to mines, factories, mills and canals. Waterways carried the engine parts for assembly on site. The Age of Steam had arrived and it drove on the process of mechanization that lay at the heart of the industrial revolution. Not least, in Birmingham.

Even with the Boulton and Watt premises on its own doorstep, it's hard to see how this transformation of Birmingham into the city it became could have happened had the Birmingham Canal Company been allowed to carry on its short-sighted and avaricious monopoly of trade. As L. T. C. Rolt put it, the company 'automatically opposed any scheme which threatened to divert traffic from travelling the greatest possible distance over its own canal and so earning maximum toll revenue ... doing nothing to improve conditions on its own main line ... despite the golden harvest it yielded it remained tortuous, narrow and shallow'. Like the behaviour of its owners.

Steam engines under construction at the Boulton and Watt Foundry at Soho, near Birmingham

Thanks to pressure exerted by the manufacturers and merchants of Birmingham, the company was not allowed to maintain its stranglehold. In 1781 the company's old adversary, Samuel Garbett, along with other business interests fed up with the Birmingham Canal Company's machinations, came up with a proposal for what would become the Birmingham and Fazeley Canal.

The original plan was for a canal that went from Wednesbury to join the Coventry Canal, a new branch of which would be built to Fazeley. The proposed canal would be, without question, a direct and serious rival.

The Birmingham Canal Company determined to try and see off these bold plans with 'all possible opposition'. It now proposed two new branches, which would be under its control: one from Farmer's Bridge to Fazeley, another from Riders Green to Broadwaters, which lay between Wednesbury and Walsall. (The rival company had already proposed such a line.) Battle was joined.

As an early historian of Birmingham put it:

Both parties beat up for volunteers in the town, to strengthen their forces; from words of acrimony, they came to those of virulence; then the powerful batteries of hand-bills, and newspapers were opened . . . every corner of the two houses [of parliament] was ransacked for a vote; the throne was the only power unsolicited.

The Birmingham company won the battle but lost the war. Its two new branches were accepted, but immediately the Act was passed, it bought out the rival scheme's shareholders, and the two companies were amalgamated in 1784. The new concern that resulted was so good they named it twice: the 'Birmingham and Birmingham and Fazeley Canal Company'. Meanwhile, work on this new and valuable addition to the Birmingham Canal network went ahead. (With John Smeaton as its engineer.) The eastern section was now to pass through Aston on its way to the Coventry Canal.

The new network was completed in 1789. By 1793 a hundred boats a day were passing along it. The next year it became known by the more manageable monicker, the 'Birmingham Canal Navigations'.

But this new spirit of cooperation didn't last long. Further attempts to enlarge the canal network around Birmingham were persistently objected to, or hobbled by the application of exorbitant tolls for cargo coming on to the BCN's own waters. A case in point was the Dudley Canal.

The Dudley Canal, 1973

This was the pet project of Lord Dudley and Ward, who owned extensive limestone quarries in the Dudley area. He proposed to build a canal that would join up with the Stourbridge Canal, which in turn would link to the Staffordshire and Worcestershire Canal. Coal, ironstone and limestone from Dudley and South Staffordshire would now make their way down to the Severn. The joint scheme gained strong support from the key transport hubs of Worcester and Gloucester, as well as the many glass-makers who would be served by it en route. This caused considerable alarm at the Paradise Street head-quarters of the BCN, who vigorously opposed the 1775 Bill and succeeded in getting it withdrawn.

However, as was so often the case where a canal that made sound economic sense (and had strong backing) was concerned, it was able to overcome the opposition of vested interests and was authorized the next year. In 1779 the Dudley and Stourbridge Canals were completed.

But Lord Dudley and Ward wasn't entirely satisfied with the business arrangements. The Staffs and Worcs proprietors, exhibiting the greed and stupidity that so frequently characterized the canal business, were taking full advantage of their control of water transport to the Severn and charging a 50 per cent premium on coal. Dudley now looked eastwards. The BCN agreed that he could build a tunnel, and then a short canal (Lord Ward's Canal) on to their main line at Tipton.

Naturally, they too charged a special toll for traffic coming from the new branch, on the grounds that it reduced the amount of tonnage coming from north of the junction. But for the time being Dudley was content to go ahead. (And, of course, the Staffs and Worcs company was forced to lower their own tolls.)

The BCN now went on a rare efficiency drive and made some important improvements to the main line. When the canal was first constructed, it operated on two separate levels. The Birmingham Level, 453 feet above sea level, and the Wolverhampton Level, 20 feet higher. Six locks were now eliminated and a cutting dug so as to lower the summit level at Smethwick. This meant that the whole section from Smethwick to Wolverhampton was lock-free, an enormous saving of time.

This had an important effect on the nature of the traffic on this section of the BCN Main Line. The fact that there were no locks to negotiate meant that boats could be tugged in strings. Most of the traffic (as elsewhere on the canals) was local, and so the majority of the journeys were undertaken by day-boats (known in Birmingham as 'Joey' boats).

In 1792, however, the Birmingham Canal Navigations Company overreached itself and built a lasting monument to what Rolt would call its 'folly and greed'. This was the Worcester Bar, a stone construction 7 feet wide and 84 feet long, as notorious as it was self-defeating.

It came about because of a rival, the Worcester and Birmingham Canal. The advantage of a direct route between the two places was strikingly clear. It would shave off several miles from the existing line that went through the Dudley Tunnel (laboriously slowly), down through the Stourbridge Canal and on to the Severn. Its undoubted utility meant, naturally, that it was frantically opposed, not only by the BCN, but also by Lord Dudley and Ward and the Stourbridge proprietors.

But then, to the spitting fury of the BCN, Lord Dudley changed sides. Fed up with the company's intransigence and high-handedness (and not happy at the tolls he was being charged), His Lordship now proposed to build a new canal down to meet the planned Worcester and Birmingham Canal at Selly Oak. (This would need the construction of another long tunnel, the Lappal, which weighed in at 3,795 yards.)

To rub salt into the wound (and indeed salt would be one of the major cargoes) his new canal would run almost in parallel with the Birmingham Main Line. Coal and other traffic would be able to come from Dudley into the heart of Birmingham without paying any tolls to the BCN whatsoever.

The aggrieved proprietors now whipped up another storm of protest and hysteria, playing the doomsday card and arguing in pamphlet after pamphlet how disastrous the new canal would be, especially on the finances of widows and orphans, who, they implied, made up the great majority of the canal's shareholders.

The No campaign failed. After a petition 14 yards long, which contained over 6,000 signatures, had been presented to parliament, an Act authorizing the Worcester and Birmingham Canal was passed in 1791.

The Birmingham Canal Navigations Company, however, refused to allow the interloper to join its hallowed waters. And so at its terminus at Gas Street Basin, Birmingham, the new canal was greeted by the Worcester Bar, which physically separated the two canals. Goods had to be hoisted by crane in and out of the boats and carried physically along the bar. The official reason for the Worcester Bar was to prevent the new canal stealing water from the old. But this fooled no one. Its purpose was quite obviously to prevent the Worcester and Birmingham from carrying cargo on the BCN's own canals. Only in 1815, following an Act of Parliament, was the Bar replaced by a stop-lock (the traditional means of regulating water between two canals).

The BCN didn't oppose another canal, the Wyrley and Essington (later known as the 'Curly Whirly'), which was authorized in 1792. The Wyrley and Essington linked Wolverhampton to the coalfields of Cannock Chase and was eventually to join the Coventry Canal, thus providing an outlet to the Trent and Mersey in one direction, the Birmingham and Fazeley in the other. The BCN, naturally, did charge special compensation tolls on coal coming on to its own network. As they had, perhaps, learned a lesson from their experience with Lord Dudley, these were kept to a moderate rate.

That's not to say that more battles weren't to be fought. Also in 1792 another challenger appeared on the horizon. This was the Stratford-upon-Avon Canal, and it was far less to the BCN Company's liking. Completed in 1793, it ran from Kings Norton, on the Worcester and Birmingham Canal, down (eventually) to Stratford-upon-Avon itself, and thus to the Severn at Tewkesbury.

Worse still, from the BCN's point of view, a further line was proposed that would branch off from the Stratford-upon-Avon Canal at Lapworth, with the intention of joining the Grand Junction Canal at Braunston. This far more direct route would neatly cut the BCN, Coventry and Oxford Canals out of the trade from Birmingham to London. The Worcester Bar now rose to bite its constructors. Because the new canals would link up with the

Worcester and Birmingham, the BCN's own bar would prevent traffic being carried from its own network down the new line to London.

The BCN opposed the plan, of course, but more sensibly sponsored another canal, the Warwick and Birmingham, which ran from the Birmingham and Fazeley line down to join the Stratford-upon-Avon Canal at Lapworth. This would be under its effective control.

The Oxford Canal, meanwhile, managed to get the proposed Warwick branch of the Stratford-upon-Avon Canal diverted to its own waters, by altering the destination of the line from Braunston, as originally proposed, to Napton. This became the Warwick and Napton Canal, and kept the Oxford Canal in the Birmingham to London loop.

In 1799, after more than twenty years of arguments and prevarications on the part of the BCN proprietors, a line from Wednesbury to Walsall, with its important coalfields, was finally built. It became the Walsall Canal.

With these new lines, with their scores of branches linking to collieries, mills, foundries, factories, workshops and wharves, the Birmingham Canal Navigations, at their height a network of 160 miles, were outstandingly profitable.

By the 1830s the industrial revolution was in full flood, and production of iron dramatically increased. Birmingham's industries fizzed and crackled. Until the railways and later road haulage took away much of their traffic, millions of tons of raw materials floated into the wharves that lined the canal banks. Coal, ironstone, limestone, copper and zinc, for iron-, brass- and steel-making. Clay, for bricks, pottery and tiles. Slate from Wales, and timber from the Baltic and North America. Sand for glass- and iron-making. From Asia gums and resins needed for veneer and lacquer. Metal for Boulton's coins. Wheat, barley, hops and other foodstuffs to be made into bread and beer. Wine, spirits and other 'luxury goods' arrived for consumption by the well-to-do.

Later, as the 'second' industrial revolution moved into a more chemical-based phase, salt, phosphates, sulphur and other ingredients were needed for works that treated metals and cloths and produced soap and other substances.

In the opposite direction, and on to Liverpool, Hull and London, went swords, guns, ammunition, brass bedsteads, trays, nails, anvils, hammers, fencing, water troughs,

Barges bringing fresh milk to the Cadbury Factory at Bournville, Birmingham

chains, pumps, hurdles, oil lamps, pots and pans ('holloware'), belts (and braces), metal jewellery, soap, candles, matches, paint, glue, acids and alkalis (both of which had a variety of final uses, industrial and agricultural), tar, pitch, creosote and later bicycles (the latter from BSA, later a manufacturer of motorbikes; it began life as Birmingham Small Arms, a maker of ordnance). Chocolate, until the company moved to Bournville in 1879, was produced by Cadbury at their site in Bridge Street. Like many other substantial firms, Cadbury owned their own fleet of boats.

Trade was brisk. Industry was using up ever greater amounts of iron, and on Birmingham canals traffic in the metal more than doubled between 1832 and 1845. Tonnage of coal, lime and limestone also increased.

But in the 1830s, railways began to crash the party. As early as 1825 a railway lobby pamphlet had suggested that over the preceding five years on the BCN an average of thirty days per year had been lost to frost, maintenance work and other closures. The message was that railways were the transport system of the future, and that the inefficient, monopolistic, greedy canals had had their day.

At first the BCN, as was its wont, fought back and gave strong opposition to the interloper. But by the mid-1840s it concluded that resistance was useless. However, this wasn't the end of the line. The canal network in Birmingham's overcrowded banks and many private cuts actually gave it a considerable advantage in keeping its head above water. It was impracticable for a railway physically to intrude near to the canalsides. So instead 'interchange basins' were built, and canals and railways worked together, goods being transferred from one to the other by road or tramroad. (The same was true in other major industrial conurbations – where it suited the railway companies' book.)

By 1845 the BCN was in the railway business itself, cooperating with the London and Birmingham and Shrewsbury and Birmingham companies to build new railway lines together. In the same year the BCN leased its canal to the London and Birmingham Railway Company in return for guaranteed dividends of 4 per cent p.a. (The Birmingham and London became part of the LNWR.)

More branch canals were built. The Tame Valley and Rushall Canals were finished in 1844 and 1847. The incorrigible tunnellers of the Dudley Canal were also at work. Starting in 1855, they made the last ever canal tunnel to be built in Britain. The Netherton Tunnel, 3,027 yards long, 27 feet wide, with double towing paths, lit by gas, alleviated the notorious bottleneck caused by its predecessor, the Dudley Tunnel.

The Birmingham Canal network had been an enormous success, a handmaiden to Birmingham's expansion – particularly in the 1830s to the 1850s – and a vital component of industrial revolution in the town. The results, however, were not pretty. In 1858, the same year the Netherton Tunnel was finished, a London journalist, John Hollingshead, visited the Birmingham Canal network by boat. He drifts:

past black, smoke-grimed, town-stamped boys, angling in the canal; past groups of ill-favoured, smoking, half-drunken men-youths . . . past a coal-dust looking towing-path, and under a sky of smoke; past tall chimneys and dingy gas-works; down another staircase of black locks, opened by a poor, active, grimy girl-child not more than ten years of age . . . past backs of myriad-windowed factories, whose glass is broken, and under whose walls lie green sickly pools of stagnant water; past a dozen grimy boys with large set jaws and shrunken arms, and legs, bathing amongst the floating dead dogs and factory scum of the inky canal.

Today, it's good to report, the canal is somewhat more salubrious.

The Birmingham Canal Navigations – the journey itself

Old Street Basin, Birmingham

At its peak in the 1850s the Birmingham Canal Navigations comprised around 160 miles of canal. Today 100 miles is navigable. To traverse all of that would take around a week, and it's probably fair to say that the further-flung branches, such as the Tame Valley and Walsall Canals, are more for the hardcore enthusiast. The two Main Lines, and their still operative branches, and the Birmingham and Fazeley Canal are the focus of our journey here.

It's often pointed out that Birmingham has more miles of canals than Venice or Amsterdam. It may not have all the charm of those two cities, but the triangle formed by Oldbury, Wednesbury and Tipton is a tangled skein of lines and branches which, just looking at it on a map, gives us an inkling of the sheer amount of industrial activity that went on here. The collieries, brick works, quarries, foundries, factories, wharves and warehouses that made Birmingham the 'City of a Thousand Trades' jostled for position on a network that saw 200 working boats pass every day. In his *Shell Book of Inland Waterways*, Hugh McKnight remarks, 'If the canal and river network of England is considered as an untidy spider's web, its most closely woven centre is the BCN.'

We start our journey, not on the Birmingham Canal Navigations proper, but on its old rival, the **Worcester and Birmingham Canal**, victim of the infamous Worcester Bar. It's hard to resist because of its dramatic entry into the city centre.

From **Worcester**, where it joins the River Severn, to **Tarde-**bigge, near Bromsgrove, there are an extraordinary 58 locks over 16 miles. The Tardebigge flight alone – the longest in England – is made up of just over half of them, in just over 2 miles, rising 220 feet. (In the early 1800s a lift was built, but it never really worked.) Though not for the faint-hearted, this is a much-celebrated section of canal.

Now lock-free, the canal passes through the Tardebigge and Shortwood Tunnels, along the peaceful countryside of the Arrow Valley and on to the large village of **Alvechurch**. From here there are views towards Weatheroak Hill, which is crossed by the old Roman road of Ryknild Street. Lickey Hills, the southern end of the South Staffordshire coalfield, can also be seen to the other side. **Bittel Reservoir** is a haven for wildlife, especially birds.

The journey then continues through the **West Hill Tunnel** (known today as 'Wast Hill Tunnel', after the pronunciation of generations of boaters). At 2,726 yards it's one of the longest still in use. The canal emerges from the tunnel to make its way to **Kings Norton**. Once a village, it is today part of the Birmingham suburbs. Not that arriving by canal would tell you this, as it still retains a bucolic feel.

The church of **St Nicolas** is noted for its fifteenth-century tower, with its octagonal spire. The Rev. W. V. Awdry, of 'Thomas the Tank Engine' fame, was a curate here in the 1940s. Near the church is the former **Grammar School**. Closed in 1875, it's been restored and is now a Scheduled Ancient Monument. Curiously, the half-timbered upper floor dates from the fifteenth century, the ground floor, in brick, from 200 years later.

At **Kings Norton Junction** the Stratford-upon-Avon Canal joins the Worcester and Birmingham, both, when first conceived, plotting to break the BCN and Oxford Canal's monopoly. There's an unusually imposing **toll-house**, of Victorian vintage.

The next point of interest is **Bournville**, the 'model' factory village which the famous Quaker firm of Cadbury began building in 1879. Founded in a shop in Bull Street, Birmingham, in 1834, the company later moved to larger premises at Bridge Street. But 'Why should an industrial area be squalid and depressing?' asked George Cadbury. The answer was the greenfield site of Bournville, eventually to contain 3,500 houses spread over 1,000 acres. The canal served the factory with cocoa until 1961. Like many larger businesses, Cadbury owned its own fleet of boats, with their distinctive red and green livery.

In the 1920s another George Cadbury served on several committees and fought to keep the canal open. The **Cadbury World** exhibition is on the site. The business, highly controversially, was taken over in 2010 by the giant American processed-cheese company Kraft.

At **Bournbrook**, Selly Oak Junction used to link with the Dudley (No. 2) Canal, which closed after the Lapal Tunnel collapsed in 1917.

Another important institution has its own moorings on the canal, at **Edgbaston**. **Birmingham University** was one of the first 'red-brick' universities, founded in 1900. Its distinctive campanile is nicknamed 'Uncle Joe' – not named in honour of Stalin by student radicals, but for Joseph Chamberlain (father of

Neville), a former mayor of Birmingham, and the first chancellor of the university. The **Barber Institute** contains an important modern art collection.

On the opposite bank, part of a Roman fort has been reconstructed. The 15-acre **Botanical Gardens** are to be found just before the short **Edgbaston Tunnel**. They were laid out in 1829 by the important garden designer and botanist J. C. Loudon.

Of the journey from Edgbaston into the city centre, Robert Aickman wrote, 'the Birmingham end of this waterway provides a striking example of the principle that canals stretch green fingers into towns'. There are several attractive old bridges overhung with foliage, which the canal passes under as it makes its way towards central Birmingham.

The new developments of the city now make their presence felt, starting with the **Cube**. Constructed on the site of a former stable, this building, with its jigsaw pattern, was so complex that no contractor would touch it, and the developers had to build it themselves.

Overlooking the final stretch of the Worcester and Birmingham, formerly Salvage Turn, is the **Mailbox**, a huge shopping, dining and office complex (so named because it's on the site of what was once the country's largest sorting office). When the Birmingham Canal was first built, a wharf was near this site – still remembered in the name of the road that runs through the Mailbox, Wharfside Street. The original company HQ was at the western end of Paradise Street, abutting the 'Old Wharf', long since paved over.

It's an undoubted shame that when the process of upgrading began many old canalside buildings were demolished in the city centre, rather than renovated for new uses. But some original features do survive, including the fine cast-iron aqueduct at Holliday Street. This is one of many iron bridges along the canal network (not just the BCN) built by the local firm Horseley Iron Works.

At **Gas Street Basin**, former site of the Worcester Bar, the canal now formally joins the **Birmingham Main Line**. This whole area has been much developed in recent years and is now a busy leisure hub, with smart restaurants, cafés, walkways and moorings along with office buildings – very different from the parlous state it had fallen into in the 1970s, when it was half-forgotten at best, an eyesore at worst. Birmingham was one of the first cities to realize that its waterways could be a valuable amenity. Several handsome nineteenth-century canal buildings survive.

At **Old Turn Junction** the BCN now splits into two, the Main Line heading west, the Birmingham and Fazeley Canal taking off for the north-east.

The Birmingham and Fazeley Canal

The line starts at Old Turn and continues to **Huddlesford Junction**, where it meets the Coventry Canal. The first section, to **Salford Junction**, was known to generations of boaters as the 'Bottom Road'. (Indeed, legend had it that it was indeed bottomless. The various detritus often thrown into it has disproved any such theory.)

It's fair to say that it was not fondly regarded by the people who had to work along it. In 1948 Emma Smith, in *Maidens' Trip* (a fictionalized memoir of her two years spent working as a volunteer boatwoman during the Second World War), reports that

The Bottom Road had cast a gloom on all our spirits. Surely no locks in England were as dirty as the flight at which we now arrived. The towpath looked and felt as though it had been laid with wet soot and the dung dropped plentifully on it by the many passing horses was no improvement. The lock walls were covered, not with that green slime natural to age and water, but with a black oily substance, as thick as treacle. Beams, paddles, everything we touched or tried not to touch, was coated with this same sticky dirt ... The factories we passed were squalid affairs with blackish bricks and smothered glass. They crouched beside that disenchanted water like old slum women nourished on gin and disease.

By the 1960s the canal had got even grimmer, as traffic on it stopped, and it became a festering wound running through the city. But, beginning in 1969, the bicentenary of the first Birmingham

Canal, improvements began to be made to the city-centre canals. The thirteen **Farmer's Bridge Locks**, which run to Aston Junction, once surrounded by dense industry, are now a chic part of town. Nearby is the historic **Jewellery Quarter**, and at **Cambrian Wharf** is a modern pub, the **Longboat**, which looks out at the pleasure craft lining the moorings.

It's not all new buildings. There are old canal cottages on **Kingston Row**, as well as a little toll-house and a restored lock-keeper's cottage to remind us of the working history.

At **Aston Junction**, where the Digbeth Branch joins up with the Warwick and Birmingham Canal, and thus the Grand Union, a cast-iron bridge built by Horseley Iron Works stands right next to an old brick roving bridge. On the Digbeth Branch were once the homes of household names – HP Sauce and Typhoo Tea.

At Salford Junction, the Tame Valley Canal, opened in 1844, takes us on to the modern monstrosity of Gravelly Hill Interchange, better known as **Spaghetti Junction**, and even less Italianate in flavour than

Birmingham's once busy canals. From here one can continue eight miles along the Tame Valley Canal to Wednesbury and back to the centre of Birmingham. For the less ambitious, it's an option simply to retrace one's steps, and travel back to Old Turn Junction, ready to tackle the original Birmingham Canal.

The Birmingham Main Line

Almost as soon as the Birmingham Canal begins at Old Turn it swoops to the west, indulging in a puzzling detour through the **Oozells Street Loop**. This U-shaped stretch is explained as soon as one reaches the recent development of **Brindley Place** with its many eateries and bars. This is the first of many of James Brindley's canal turns.

Meanwhile Telford's New Main Line carries on in a straight path from the beginning of the Icknield Port Loop. The modern developments peter out once we pass the **National Indoor Arena**, opened in 1991. After several derelict factories, we come to the next Telford improvement.

Brindley's original line ran in a sharply compressed curve

south-west to the Rotton Park (now 'Edgbaston') Reservoir and then north by north-west to Soho, in another (admittedly, shallower) case of the bends. Telford's New Main Line cuts straight through these meanderings, leaving them as the **Icknield Port** and **Soho** Loops respectively (both of which continued to be used to serve the businesses built alongside).

Today, the Icknield Port Loop is surrounded almost completely by waste ground, while the Soho Loop has few buildings of much interest. Instead, one must imagine what this stretch would have been like 100 or 200 years ago, when it was a buzzing hive of industry. This is true most of all at the far end of what is now **Hockley Port**, once the entrance to the world's largest industrial complex, Boulton and Watt's Manufactory, where their revolutionary steam engines were built. (Boulton's Mint was also here.)

Soho Foundry, built by Boulton and Watt in 1795, is further along the Main Line, and now home to the weighing machine makers Avery Weigh-Tronix. This was another giant, and the first factory to be lit by gas (thanks to William Murdock), so the workers could carry on into the night. The loop that served it is now dry.

After a stretch made up largely of waste ground, abandoned buildings and railway and road bridges, the canal reaches **Smethwick Junction**, where the Telford New Main Line (this section is known as the 'Island Line') leaves Brindley's original route, the Old Main Line.

Neither line offers too much visual interest at first, as the land either side is either derelict – which does have the advantage of turning it into something of a de facto nature reserve – or lined with functional modern industrial buildings. At the third of the **Smethwick Locks** on the Old Main Line one of the many former toll-houses has been reconstructed – not altogether accurately. Just past this top lock is a bridge, passing under which leads to Telford's separate branch, the **Engine Arm**, with its fine Gothic iron aqueduct (built by the Horseley company, based in nearby Tipton). It was named after the Boulton and Watt engine, built in 1779, that pumped water up to the top of the flight of locks on the Old Main Line from Rotton Park Reservoir. The old pumping house at Brasshouse Lane, which took over this duty at the end of the nineteenth century, is now the **Galton Valley Canal Museum**. From here both canals manage to surround themselves with some protective greenery, as they head towards Oldbury.

Another celebrated feature of the New Main Line is Telford's wonderful **Galton Bridge**. A 'lacy affair', as Hugh McKnight calls it, built in 1829 with an elegant single span of 150 feet. Today its effect, while still powerful, is lessened – as is so much of the BCN – by the presence of lowering road bridges crossing the canals.

The **Steward Aqueduct** carries the Old Main Line over the New, as the former takes a precipitous turn south-west to Oldbury. The Old Main Line now passes right underneath the M5 towards the Dudley Canal, through terrain of little interest for the sightseer.

Meanwhile Brindley's old branch line to Wednesbury passes down through the three **Spon Lane Locks**, among the oldest in the entire canal network. At the top lock is a split bridge, allowing the horse to cross from one side of the canal to the other without the tow rope needing to be untied.

At **Pudding Green Junction**, Wednesbury Old Canal leaves the New Main Line and, in part following Brindley's earlier route, travels past Ryders Green to a bewildering tangle of small branches that once served the collieries around Wednesbury, heading also to the Tame Valley and Walsall Canals.

The New Main Line doesn't fare much better than the Old as far as points of interest are concerned (except, in both cases, ones that are long vanished). Having passed by the **Gower Branch** (which links up to the Old Main Line), at **Albion Junction**, and the entrance to the **Netherton Tunnel** (which runs under the Old Main Line, carried over it by the **Tividale Aqueduct**) the new line arrives at **Factory Junction** (named for the former Borax soap factory). Nearby is the site of a former boatmen's mission. From Factory Junction we now take a detour, turning sharply south-east, along a branch that leads to the Dudley Tunnel, which joins Brindley's Old Main Line at Tipton Junction. The short stretch to the tunnel is Lord Ward's Canal, named after the man who built both, Lord Dudley and Ward.

Anthony Burton and Derek Pratt, in their *The Anatomy of the Canals – The Early Years*, describe the Dudley Tunnel as 'one of the most extraordinary achievements of the canal age, the creation of a watery labyrinth with over three miles of tunnels ... certainly the most remarkable canal tunnel in Britain'. Great engineering feat though it was, with 3,172 yards of tunnel (albeit discontinuous) hewed out of a series of caverns of limestone, it was, however, extremely small, allowing boats only to travel in one direction at a time. Finished in 1791, it was the last tunnel to use leggers, as steam tugs could not pass through. It became, of course, a notorious bottleneck – leading to the construction of the Netherton Tunnel to the east, sixty or more years later. Today no powered craft are allowed in the Dudley Tunnel, because of fumes. A battery-powered narrow boat service, however, offers passengers the chance to travel through this spectacular, atmospheric, eerie waterway.

Right next to the Dudley Tunnel is the **Black Country Living Museum**, a large open-air 'reconstructed industrial-era landscape'. A vertical lift bridge, a relic from one of the many canal–railway interchange basins, is one of the exhibits. The museum gives access to Lord Ward's Canal.

With the exception of the **Coseley Tunnel**, which cuts through yet another of Brindley's generous curves, the Wednesbury Oak Loop, the canal now continues in reasonably straightforward fashion (apart from a strange aberration near Bilston) on to Wolverhampton, lock-free. Several of the original cast-iron BCN boundary posts line the canal. From Wolverhampton to Aldersley, where it meets the Staffordshire and Worcestershire Canal, the windlass is once again required, as there are twenty-two locks (originally twenty-one, but one very deep lock was later made into two).

Before the Birmingham Main Line was finished its principal engineer, James Brindley, had died, in 1772. By then this remarkable individual had completed the Bridgewater and Staffordshire and Worcestershire Canals and laid out and supervised the other canals that made up the 'Grand Cross' – the Trent and Mersey, Coventry, and Oxford Canals (which remained unfinished at the time of his death).

And that was only a fraction of Brindley's workload. As his biographer Samuel Smiles admiringly put it:

There was scarcely a design of a canal navigation set on foot throughout the kingdom during the later years of his life, on which he was not consulted, and the plans of which he did not entirely make, revise, or improve.

He wasn't kidding. Brindley also worked on the Droitwich Canal (completed during his lifetime), as well as the Andover, Chesterfield, Forth and Clyde, Lancaster, Leeds and Liverpool, Salisbury and Southampton Canals. He had also advised the Corporation of London on improvements to the Thames (his solution was, of course, a canal), and that of Liverpool on how to keep its dock free of mud. He also offered consultations on drainage in the Fens. All these projects, Smiles helpfully reminds us, 'fully occupied the attention of Brindley'.

Future historians marvelled at how Brindley managed adequately to keep abreast of all these undertakings. His employers had asked themselves the same question. Many pointed letters were sent to him inquiring when he would next distinguish their canal with his presence. The Coventry Canal Company actually fired him, while he resigned from his role at the Oxford Company, though things were later patched up.

The fact is that Brindley was undoubtedly spreading himself too thinly, spinning too many plates, frantically racing from one project to another – or racing as far as was possible given that he would have had to travel for the most part by road, accompanied by his faithful mare, to whom he was devoted.

Wedgwood had written to his partner, Bentley, as early as 1767:

I am afraid Brindley is endeavouring to do too much, and that he will leave us before his vast designs are accomplished. He is so incessantly harassed on every side that he hath no rest for either mind or body, and will not be prevailed upon to take proper care of his health.

The next year, he wrote that his friend was:

a real sufferer for the benefit of the public. He may get a few thousands, but what does he give in exchange? His health, and I fear his life too, unless he grows wiser and takes the advice of his friends before it is too late.

Wedgwood was right to be fearful. James Brindley died at the age of fifty-five, killed by a combination of overwork and diabetes. The disease had been undiagnosed until Erasmus Darwin identified it late in Brindley's life. He refused to slacken his pace.

Brindley's legacy is threefold. Firstly, his was the vision of the Grand Cross. At the time of his death he had been responsible for the laying-out, if not the completion, of 365 miles of canal. He was clearly a remarkably broad-thinking individual, who laid the groundwork for all the canals that followed his own, even if different approaches would be taken by later engineers, especially Jessop and Telford. He was a pioneer. And if some of his methods were, with the huge benefit of hindsight, superseded, that is understandable.

James Brindley (1716–72), engineer

But two of his techniques did become problematic, albeit long after his death.

Thanks to the Harecastle Tunnel the gauge of the middle section of the Trent and Mersey was narrow, its locks accommodating boats no wider than seven feet. Of course, with the technology then available, Brindley could hardly have made a tunnel twice as wide. But the unfortunate fact remained that the Trent and Mersey's middle section, and all the other Brindley canals that joined on to it – the Staffs and Worcs, Coventry, Oxford and BCN, were narrow too. Yet they were surrounded by broad canals, such as the Bridgewater, Aire and Calder, Leeds and Liverpool, Grand Junction, Thames and Severn, and Kennet and Avon Canals. The carriers on these waterways naturally tended to favour wider boats, with bigger payloads. There were also sailing vessels on the estuaries and navigations.

So in order to transfer to the narrower canals, port facilities needed to be built so as to 'trans-ship' and transfer cargo from one boat to another. This, of course, was time-consuming and costly. Not such a disadvantage when canals were such a novelty, the fact that boats had to unload their cargo on to smaller vessels did become a considerable problem after competition from railways began.

(It's a considerable irony that, had the Weaver Navigation's mischievous scheme to miss out the Potteries and avoid having to cut a tunnel through the Peak District ridge been adopted, the narrow canal may not have come into existence.)

The most notorious aspect of Brindley's canal constructions, though, was that they were 'contour canals'. The reason was partly cost. Locks were expensive to construct, so if a canal could follow the natural level of the surrounding terrain as far as possible, they could be minimized. What's more, Brindley had a great liking for water that behaved itself.

As Smiles puts it:

He likened water in a river flowing down a declivity, to a furious giant running along and overturning everything; whereas (said he) 'if you lay the giant flat upon his back, he loses all his force, and becomes completely passive, whatever his size may be'.

Brindley had worked on the Calder and Hebble Navigation not long after a disastrous flood – he knew how dangerous and unmanageable fast-flowing water could be. Which is why, on a famous occasion when he was asked what the purpose of navigable rivers was, he replied, 'To make canal navigations, to be sure.'

But while contour canals did indeed save on expenditure on locks, deep cuttings and embankments, they had one profoundly negative consequence. Brindley's canals are meandering. Languorously they saunter their way around objects that later engineers would have been itching to tunnel through or bridge over.

The most notorious example of this is the Oxford Canal. Of one stretch it was said that a boatman could travel all day and still be in hearing distance of the church bells of Brinklow. (The same was supposed to be true of Braunston church.) The serpentine curves around Wormleighton were a particular source of frustration. (In Brindley's defence, this may have been because Earl Spencer, ancestor of Princess Diana, insisted on this. Perhaps, one theory goes, he wanted to make sure the canal served all his tenant farmers. Manure was usually carried free.) The Birmingham Main Line was another case in point.

When speed became ever more important – thanks to increased competition from other canals, and later the railways and roads – the contour canals at the heart of the network came to be a severe disadvantage.

All this was for later. At first the canals were an astounding success. The economic benefits were striking. A horse could carry 2 tons (on a good road) and pull around 30 tons on a navigable river. On a canal it could pull 50 tons. The savings over road transport were dramatic. Goods carried from Liverpool to Etruria by road cost £2 10s per ton. By canal they cost 13s 4d. From Manchester to Birmingham goods by road cost £4, by canal £1 10s. Liverpool to Wolverhampton or Birmingham by road, £5, by canal £1 5s.

In the Potteries, between 1760 and 1785, according to Josiah Wedgwood, the workforce more than doubled, from 7,000 to 15,000. When the preacher John Wesley returned to Staffordshire in 1781 (he had been given a rough reception on his first visit, and had been far from impressed) he found that 'the wilderness is literally become a fruitful field'.

It wasn't just industrial areas that benefitted from canals. Thomas Pennant, in *The Journey from Chester to London*, published in 1782, points out that:

The fields, which before were barren, are now drained and, by the assistance of manure, conveyed on the canal toll-free, are clothed with a beautiful verdure. Places which rarely knew the use of coal, are plentifully supplied with that essential article upon reasonable terms. It affords a conveyance of corn unknown to past ages.

'By the conclusion of this project,' he adds, the canal builders 'have contributed to the good of their country, and acquired wealth for themselves and posterity.'

The Reverend Stebbing Shaw, in his *Tour to the West of England*, published in 1789, pursues a similar theme. Of the Trent and Mersey, he says:

in a few years after it was finished, I saw the smile of hope brighten every countenance, the value of manufactures arise in the most unthought of places; new buildings and new streets spring up in many parts of Staffordshire, where it passes; the poor no longer starving on the bread of poverty.

He continues, 'and the rich grow greatly richer'.

As for Samuel Smiles, good Victorian reformer that he was, there was a moral dividend. The joint action of the Duke of Bridgewater, Josiah Wedgwood and their fellow promoters 'was not only to employ, but actually to civilize the people'. After the arrival of its canal, the Potteries had been transformed 'from a half-savage, thinly peopled district' into one that was, only twenty-five years later, 'abundantly employed, prosperous, and comfortable'.

This multiplier effect on money and manners was to be seen, in full measure, with the development of the first main waterway system of Yorkshire.

The Aire and Calder Navigation

Introduction by John Sergeant

When we began, in the bustling, no-nonsense port of Hull, it was quickly apparent this would be unlike any of my other canal journeys. No cream teas and quaint little pubs, it was straight into a world of hard hats and fork-lift trucks. For more than 300 years, the Aire and Calder Navigation has been one of the country's vital arteries. It is Britain's oldest canal still in commercial use.

It has certainly seen better days; there were long periods when ours was the only boat to be seen. But our trip turned into one of the most interesting and at times invigorating ones in this series. We started in style, leaving port on a large, oil-carrying barge, the last of its kind specially made to take cargoes down the Humber estuary and into the canal system. I was particularly taken by the way the wheelhouse can be dramatically lowered to allow the barge to squeeze under the bridges.

This is a splendid way to move a freight-load of oil, which would otherwise have been carried in twenty road tankers; and using the same engine power as just one of those trucks. Soon after we had steamed many feet below the beautiful Humber suspension bridge we came to the beginning of the navigation. It is technically different from a canal because it closely follows a river system, made up in this case of two rivers, the Aire and the Calder. And keeping the old name gives the navigation a distinctive label it richly deserves.

For most of its length, this is a superior canal, much wider than those designed for narrow boats. It has always been driven by ambition, constantly being improved as the demand for its services grew. Originally it was used to transport wool from the sheep farms around Leeds, but soon it was the movement of coal from the Yorkshire pits that dominated the canal traffic.

The demise of the British coal industry and the increasing use of road transport had a devastating effect on the fortunes of the waterway. Over the past twenty-five years the canal has been forced to rest on its laurels. But it is well looked after; ships from the Continent are still frequent visitors to Goole, the furthest inland of any port in the country; and in places strenuous efforts are being made to overcome the environmental damage caused by years of coal extraction. Slag heaps are being covered in trees to form parks, and I met a cheerful group of volunteers planting flower gardens at a lock.

Businesses are being encouraged to think of new ways to bring the canal back to its old self. We went to a glass factory in Knottingley, one of three in the town, which only exist here because of the canal; it brought coal for their furnaces and sand from the River Humber. The company still ships raw material down the navigation as far as Goole, and they are now considering bringing it all the way, right to their doorstep.

Maintaining some of the old craft skills is vital. One of the most inspiring small factories on the route is run by the Canal and River Trust. Here they make all the new lock gates in Britain, and each one has to be tailor made from traditional oak timber. Lord Nelson would have approved. The gates only last for about twenty-five years, and it is a sobering thought that without this factory the whole British canal system would soon seize up.

As we travelled to the outskirts of Leeds, first by barge and then by narrow boat, I was struck by how much potential this canal has to provide work and enormous amounts of pleasure. It needs investment and imagination, but at least we are long past the stage when people thought of filling in these old waterways and turning them into roads. The long-term future of the marvellous Aire and Calder Navigation looks bright.

The Aire and Calder Navigation is a striking example of an early and highly successful attempt to integrate the new canal technology with existing river systems.

Many rivers in Britain were naturally navigable – which is to say they were navigable for the purposes of trade – such as large stretches of the Thames, Severn and Yorkshire Ouse. The fast-flowing water scoured the river beds. Tidal water pushed fresh water further upstream. So, at certain times of the year, some rivers were navigable quite far inland (in the case of the Severn, for example, sometimes as far upstream as Worcester). They thus played a vital role in the country's economy. Roman roads were built with military, rather than goods transport in mind. Where possible, the Romans carried goods by sea, or by river.

But passage on the rivers was uncertain. The vagaries of weather took their toll, leaving stretches of many rivers too dry to navigate during the summer and too fast-flowing in the winter, meaning that boats could sometimes be left stranded a week or more before proceeding to their final destinations.

From the medieval period rivers became virtually impassable at any time of year. Millers built dams so as to supply the water power for their mills. 'Fish weirs' were slung across the rivers, and boats were unable to pass except by time-consuming and often expensive negotiations with the millers and weir-owners concerned, who had to be persuaded, cajoled, bribed or threatened so as to supply the necessary water, or take down the weirs, so that a boat could carry on its journey. Farmers also tapped rivers to irrigate their fields.

A pound lock

Early attempts to make rivers easier to navigate involved the building, not only of weirs and dams, but what were called 'flash-locks'. These were essentially weirs with a movable central portion, first in use around the end of the thirteenth century. They allowed a boat to slide down a wooden slope from one level to another. To facilitate this, a 'flash' of water was released from further upstream (often by a miller, whose palm had to be greased before doing so). Going downstream could be dangerous, as well as being wasteful of water – at an accident at Goring, on the Thames, in 1634, sixty people died. Going upstream meant that the boat had to be laboriously manhandled over the lock. Despite their many drawbacks flash-locks carried on being used until the twentieth century. The last flash-lock on the Thames closed for business in 1937.

A great advance came with the invention of 'pound-locks', the 'pound' being the body of water between two locks. Their use dates from at least 1577, when, on the River Lea, in east London, we have a description of 'double doores' being built. They were introduced on the Thames in the seventeenth century and began to be used in great numbers after their appearance on the Newry Canal in the 1730s. Pound-locks were able to drop to a far deeper level than flash-locks and, as well as being safer, were far more efficient in keeping water, in that only a lock's worth was sent down river on each operation.

As well as building weirs and locks, 'cuts' were made to replace – in transport terms – parts of the river that were particularly undulous or tricky to navigate. One of the first such substantial cuts to be dug was the Exeter Ship Canal, a 1¾-mile canal which was fashioned in the 1560s. Not because the Exe was unnavigable – far from it. Local aristocrats had, more than two centuries earlier, built weirs across the river to power their mills, thus blocking river access from the estuary to the centre of Exeter. And were charging exorbitant fees to allow the transport of goods into the city.

In the next century, however, a movement began to make 'improvements' to rivers that weren't navigable by nature. Sir Richard Weston tamed the River Wey (which joins the Thames at Weybridge) in the 1650s, while Andrew Yarrington canalized part of the River Stour, which flows into the Severn estuary. In 1677, this early visionary wrote a book called *England's Improvement by Sea and Land*, advocating alterations to the Severn and Trent.

In the following century this call to arms was taken up, and the Trent, Yorkshire Derwent, Mersey, Irwell, Weaver and Bristol Avon were all 'improved'. An Act of Parliament was needed before alterations to a river could be made. The people who then effectively owned the rights to carry trade on the river were known as 'undertakers'.

But when it came to navigations being made, millers proved once again to be the number one enemy, fearing – often with good reason – that their water supply would be adversely affected. Landowners objected because they suspected – again, often with good reason – that their lands would be flooded or their water meadows drained of water. What's more, farmers, indeed anyone with a local commercial interest, were opposed to any innovation that might help bring in goods from elsewhere and undercut their prices.

There were also rivalries between local towns – such as Nottingham and Burton – leading to deliberate acts of sabotage aimed at blocking a rival's water trade. The people of Nottingham also vehemently opposed attempts by the people of Derby to make the Derwent navigable, for this same reason.

Before the navigation mania there were around 685 miles of navigable river in Britain. By 1724 there were 1,160. But they were far from perfect – and were long to remain that way.

In their early days navigations rarely had towpaths, and boats were 'bow-hauled' by groups of men stumbling along as best they could. River towpaths, where they did exist, were also notorious for the gates, and even stiles, that were placed in a horse's way, often leading to injury. They might also switch at a moment's notice from one side of the river to another, meaning that horses had to be taken over by boat, or ferry. (Constable's famous painting of Flatford Mill shows a lighter that would have been used for this purpose, on the River Stour.)

Flatford Mill by John Constable, 1816–17

The Aire and Calder Navigation, authorized in 1699, was part of this important and rapid expansion of water transport that preceded the introduction of canals. The impetus giving rise to its birth came from the all-important woollen industry.

In the Middle Ages the export of wool was Britain's most important trade. It was sent to Calais (then controlled by the English crown) and then woven into cloth by European craftsmen and sold at the various fairs, such as the famous one at Champagne. But gradually the British worked out how to finish their own wool, linen and lace, and after this the products were exported ready-made (much to the fury of the Italians, French and Flemish, from whom these secrets had in some cases been stolen, in an early form of industrial espionage).

The woollen industry was particularly strong in the West Country, East Anglia and West Yorkshire. (The great 'Wool Churches' of East Anglia bear testimony to the money to be made.) So vital was wool to the economy of Britain that Charles II passed a law making it mandatory that people had to be buried in a woollen shroud – an early form of death duty.

The woollen industry, like practically all trades before the end of the eighteenth century, was 'domestic' – what used to be called, a little misleadingly, a 'cottage industry'. Although a good deal of the work did indeed take place in cottages, in earlier days. It was usual for those engaged on the land to combine farming with other economic activity. But as the industry grew in importance, specialists developed who worked at home, or in small factories, spinning wool into yarn, or weaving yarn into cloth, full time. (Women who did the former were known as 'spinsters', the name later coming to mean a woman who was unmarried.)

For cloth to be made, the raw wool had to be cleaned, combed and 'carded' into a workable roll. It was then spun into yarn and woven, on a loom, into cloth. The cloth then had to be fouled, the process whereby impurities are pounded out, often by walking on the material, giving us the surnames Fuller, Tucker and Walker. (In Wales, a 'fulling mill' was called a 'pandy', hence the recurrence of that word in many Welsh place names.) The cloth was then washed, stretched, bleached, dressed and dyed.

Power loom weaving in a cotton mill, 1835

It was the attempts to introduce mechanization to these processes – particularly in the manufacture of cotton – that were to be one of the main driving forces of the industrial revolution. The wool industry itself was slow to take up the new machines that were to develop. But it was certainly not a small-scale industry. The processes involved in the making of woollen cloth needed considerable capital outlay. Gradually the industry came increasingly to be dominated by 'clothiers', wealthy middle men who could afford to stump up the money to buy the wool, buy and loan out the machines, pay for the finishing processes and commit to their products' export. It was clothiers who were the main force behind the Aire and Calder.

As was the case with the Bridgewater, Trent and Mersey, and Birmingham Canals, the problem that stimulated the early promoters was, of course, that of transport. The local wool of the West Riding was rather coarse, meaning that only the rougher 'worsted' cloth could be produced. Finer wool had to be imported, from Lincolnshire or Leicestershire. And here was the problem.

Before 1699 the wool arrived by road, while the finished cloth was sent from Leeds or Wakefield, also by road, to the nearest points on the navigable rivers. Coal took an even more elaborate route. Even though the coalfields of South Yorkshire were only 20 miles away, the lack of adequate road transport meant that coal had to be carried by ship from Newcastle to the Humber, and then up the Ouse to Selby – a trip of 200 miles.

The scheme for the Aire and Calder Navigation was put into motion by the clothiers of Leeds, headed by the mayor, and supported by various unnamed 'Gentlemen of Wakefield'. The aim was to link these two towns (as they then were) to the Humber. The Aire was to be made navigable from Leeds to Weeland. From here the river was already navigable down to Airmyn, where it joined up with the Yorkshire Ouse, thus giving access to the Humber estuary and the important port of Hull. The Calder was to be made navigable from Wakefield to Castleford, where it meets the Aire.

The petition for the navigation, drawn up in 1698, made clear:

> *That it will be a great improvement of trade to all the trading towns of the north by reason of the convenience of water carriage, for want of which the petitioners send their goods twenty-two miles by land carriage, the expence whereof is not only verey chargeable, but they are forced to stay two months sometimes while the roads are passable to market.*

The Bill for the Aire and Calder Navigation was supported far and wide – except in York. That city controlled all trade on the Yorkshire Ouse and argued that the proposed alterations to the Aire would drain the river and destroy their trade along it. Of course, they also feared the competition.

Trinity House was called in to investigate and ruled that the Ouse would not be adversely affected by the projected navigation. The scheme went ahead in 1699. Commissioners were appointed to protect the interests of landowners, and cuts, weirs, locks and towpaths were built. All but a mile or so of the Aire section was finished by 1701, the Calder branch by 1702. The whole navigation was open in 1709.

After complaints from landowners that the river was now prone to flooding and silting, compensation payments were made. Seeing off these obstacles, and carrying mostly wool and coal, the Aire and Calder Navigation was a financial success, and from 1718 dividends were paid to the undertakers, starting at 7 per cent. Corn and grain from Lincolnshire would also become an important source of cargo.

The Aire and Calder was unusual in that from the start of its operations it also acted as a carrying company, the boats being owned by the navigation's lessees. In 1744 Airmyn was developed into a trans-shipment port.

The adjoining Calder and Hebble Navigation was begun in 1759 and completed in 1770, running from Wakefield down to Sowerby Bridge, west of Halifax. The first engineer in charge was John Smeaton.

The Aire and Calder Navigation through Leeds, 1828

Smeaton, born in 1724, first coined the term 'civil engineer', as opposed to military engineer, and founded the Society of Civil Engineers in 1771. A scientist as well as a civil engineer, he had been elected to the Royal Society in

1753, and later won the prestigious Copley Medal for his important contribution to the improvement of the efficiency of windmills and waterwheels. His most famous work was the Eddystone Lighthouse, near Plymouth, completed in 1759. Smeaton pioneered the use of the remarkable hydraulic powers of lime, which can set under water. He used it on the lighthouse and the masonry for the Calder and Hebble's locks. His work led to the development of Portland cement.

James Brindley was also employed on the Calder and Hebble, for a short while, thus making a common link between the two most important canal builders of the early phase.

The drawback with the Calder and Hebble Navigation was that its locks were built to accommodate 'Yorkshire keels', which were only 57 feet in length. So narrow boats, 14 or so feet longer, couldn't use the waterway. This was one of the many anomalies of the canal system. It may not have seemed important at the time, but it became a considerable drawback once the canals, with their casually disparate dimensions, came to face up to the threat of the railways.

Even so, the Calder and Hebble played an important part in the Aire and Calder's trade. Wool could now be transported from Halifax, Wakefield and Leeds down to Airmyn, and then on to Hull. It then went as far afield as Germany, Holland and the Baltic.

The undertakers, happy to pocket their profits, didn't see too much need to shell out for improvements. The Aire was controlled by just one lessee, and thanks to his inactivity in terms of maintaining it the river was becoming increasingly unfit for trade. And so, as would often be the case in British waterway history, competing concerns began to stir themselves to break the monopoly, especially when a boom in West Yorkshire cloth-making developed around 1770. Rival canals were now proposed.

The Aire and Calder proprietors took decisive action. John Smeaton was called in and in 1772 made recommendations for an extensive series of improvements, including a canal. An Act was passed in 1774, defeating the other petitions, and Smeaton's assistant, William Jessop, was appointed to engineer a canal from Haddesley, on the Aire, to Selby, on the River Ouse.

John Smeaton (1724–92), engineer

William Jessop was to go on to become one of the most important of all canal builders (and much else besides), being responsible for the Grand Junction Canal, as well as supervising Telford's work on the Llangollen and Caledonian Canals. He was born in 1745. His father worked on the Eddystone Lighthouse and became a friend of John Smeaton. Smeaton took the young man on as an apprentice and became his guardian. Jessop became the secretary of Smeaton's Society of Civil Engineers.

The Selby Canal was finished in 1778, and Selby took over from Airmyn as the principal inland port. Here cargoes were loaded into larger vessels and taken down the Ouse to the Humber estuary.

By 1785 most of Smeaton's other suggested improvements had been completed, and all the old locks replaced. The company was vigorous in buying up mills, and land where it could build wharves. Stone, lime, corn and coal were now the main categories, but half of the total carried was 'miscellaneous goods', which included bales of cloth, groceries and wool.

Trade was increased as more canals joined the network. In 1802 the Barnsley Canal was finished and joined the Calder branch near Wakefield. In 1804 the Rochdale Canal opened from Manchester to Sowerby Bridge, where it met the Calder and Hebble. The Huddersfield Narrow Canal, finished in 1811, also joined the Calder and Hebble.

During the Napoleonic Wars not much had been done by way of developments to the Aire and Calder (except for the appearance of an experimental steam passenger boat, in 1813). After Waterloo, rival canals were again suggested. Once again the company acted with determination.

In 1816 steam packets were laid on as a regular service, carrying not only passengers but also shipments of wool.

The waterway was enlarged to enable it to carry larger vessels and speed up their journey. Some of the locks on the upper reaches were (again) replaced by larger models. But the Selby Canal, at five feet deep, was too shallow, and the lower reaches of the Aire were difficult to keep navigable, needing constant dredging. Selby had wharves but no dock to deal with the larger vessels.

John Rennie, who had engineered the Kennet and Avon Canal, was consulted, and he suggested a new canal, to run from Knottingley down to Goole, thus replacing the old tidal section of the River Aire. The scheme was adopted, and an Act obtained. Goole now became a company town, run by the Aire and Calder Company until 1875.

Later it was decided to extend the canal westwards to Ferrybridge (though it was still referred to as the Knottingley–Goole Canal). It was finished in 1826, weighing in at 18 miles long. A passenger service was instituted, including that offered by the company's own steam packet, the *Eagle*.

The new canal was a considerable improvement on the old navigation and led to the port of Goole replacing that of Selby in importance, and even challenging the mighty Hull. Though at first it proved inadequate for sea-going trade, and most of the coastal traffic in coal for London remained in the hands of Newcastle, Goole supplied the nearer ports.

The management of the Aire and Calder had seen the future: railways. In considerable contrast to that of many other canals, the committee was forward-thinking, determined and efficient, with an eye always to improving its navigation, and thus its business. It could see that the only way to take on the railways was through steam. And that would require bigger canals and stronger banks. This it set about to achieve.

In 1831 the Aire and Calder introduced a steam paddle tug between Leeds and Goole. The experiment worked, and more steam tugs were built. By 1835 the Aire itself had been upgraded, by a series of canalizations, all the way from Hunslet, in Leeds, to Allerton Bywater, just above Castleford.

These improvements were engineered by Thomas Telford, later to work on the Llangollen and Caledonian Canals. The upgraded navigation was 10 miles long, 7 feet deep, with locks 18 feet wide, strong enough to cope with wash from the steam tugs. Meanwhile a new canal was also dug in place of the Calder Navigation, from Wakefield to Castleford, cutting five miles off the old river route. This was overseen by George Leather.

In 1834 the railways made their first baleful appearance in the area, with the opening of the Leeds and Selby Railway. Railways would, eventually, clobber the navigation's own passenger service. But thanks to the vigilance and effectiveness of its management, the Aire and Calder was able to withstand the threat to its commercial trade.

It also expanded. In 1854 the Aire and Calder Company took over the Barnsley Canal, in which it had long had an interest. This would eventually give it a connection to Sheffield, and its famous steel.

The committee did hit the panic button once and offer

to collaborate with the railways in setting tolls. But in 1855 they chose the path of defiance and decided to remain independent.

The screw propeller took over from paddles in the 1850s, and steam tugs which carried cargo as well as towing strings of other boats were introduced. They could pull up to six or seven other vessels. When the locks were enlarged yet again they could take as many as thirteen in one go.

By 1855 two-thirds of the navigation was served by steam power. In 1857 use of the company's tugs was offered free to 'bye-traders', or independents, from Wakefield to Goole, to speed up traffic. Such a sensible course of action is hard to imagine happening in the case of some of the navigation's more short-sighted cousins.

Goole in the 1900s

The Aire and Calder committee also tried to unite with the other canal companies to take on the railways and were behind important legislation to aid canal carrying, and more flexible tolls. They were the driving force behind the Canal Association, formed in 1855. But their pusillanimous counterparts were considerably less doughty in the fight to keep the canals going. The directors of the Calder and Hebble Navigation had also started to wobble, and in 1865 it was leased by the Aire and Calder, who were determined to keep it in business.

In 1865 the most dramatic innovation of all made its appearance. W. H. Bartholomew, the able son of a manager of the Aire and Calder, and now the chief engineer, developed the 'compartment boat'. It worked on the same principle as that of the Duke of Bridgewater's starvationers and the 'tub-boats' of the Shrewsbury Canal, being a form of 'containerization'. The boats were equipped with sets of chains that linked them to the tug, and they were steered from the back of the articulated convoy, by tightening or loosening the chains. (Towards the end of the century, spring buffers were added, and the boats were steered from the front. Eventually the locks were built big enough to take nineteen boats at one time.) The compartments, once they arrived at the large dock at Goole, were lifted out of the water by a specially built hydraulic hoist, also of Bartholomew's design, and tipped into the waiting coal ship. The development of these craft, known locally as 'Tom Puddings', allowed the Aire and Calder to keep its vital coal trade out of the clutches of the railways.

And coal – always the great staple of most canals – was

of even more commercial importance now that American grain imports at Liverpool were beginning to affect British agriculture, sending it into a long depression, and the corn trade to Leeds and Wakefield was withering away. (The old wool and cloth trade, to serve which the navigation had originally been built, had for the most part gone to the railways.)

In the mid-1880s the Selby Canal was finally deepened to seven feet, standardizing it with the rest of the system. In 1905 the Aire and Calder, with the Sheffield and South Yorkshire Navigation, completed the building of the New Junction Canal, which linked to the River Don Navigation, and thus Sheffield. (The Dearne and Dove Canal, finished in 1804, offered a route to Sheffield from Wakefield, as it joined the Barnsley Canal to the River Don near Rotherham.)

Although, like all canals, the Aire and Calder suffered after the First World War, it continued to be commercially viable right down to the present day. When the demand for coal declined in the 1960s, petroleum took over as the canal's main cargo. As late as 1967 the Leeds to Goole line was enlarged so that it could take barges of 500 tons, some twenty times the amount carried by a narrow boat. Trade in general merchandise was also encouraged and continued, alongside tar, sludge, sand and paper, as well as coal for the giant Ferrybridge 'C' power station.

The forward thinking, watchfulness, wisdom and energy of its management had kept the Aire and Calder a going concern. But the real key to its success was that it linked up to the Humber, and thus global shipping. This is

1. (*top*) Boats moored near the overflow weir at Anderton. 2. (*above*) Southern end of the Harecastle Tunnel.

3. (*opposite*) Geese by the Birmingham Canal. 4. The Brindley Place development in Birmingham.

5. (*opposite*) Cruising towards Burnley. 6. Bournville Station overbridge.

7. (*top*) The glorious Pontcysyllte Aqueduct. 8. (*above*) View of Ben Nevis.

9. Reflections on the Aire and Calder canal.

10. Titus Salt's Factory in Bradford.

The Aire and Calder navigation –
the journey itself

Woodlesford Lock

The Aire and Calder Navigation is unique amongst our eight canals in that commercial traffic still uses its water – large barges carrying oil and aggregate cargoes. From a strictly leisure point of view, though, the navigation is a victim of its own success. Its constant upgrading and resizing kept it in pace with modern industry until very recent times – when power stations finally stopped using local coal. This, coupled with the rapid decline of industry in these parts in the last thirty years, means that the canal, with the exception of the chic developments of central Leeds, offers fewer spectacular stretches of scenery or historic buildings than do our other journeys. It is, however, a testament to the economic success that could – perhaps – have been within the reach of more of our canals, had a similar will been there to keep them abreast of the times. Old photographs and illustrations really bring the canal to life. The navigation from Goole to Leeds is 34 miles long, with only thirteen locks to negotiate.

Our journey starts at **Goole Reach**, where ships arrive from the North Sea, via the Humber, and then down the Ouse, in order to trans-ship cargoes for the Aire and Calder Navigation, and thus destinations inland. Goole was a town built and owned by the canal company. Construction began in 1822. By 1841 its population had increased sixfold. Not until 1933 did the town become incorporated as a borough. Goole is 45 miles from the North Sea, Britain's most inland port.

Goole's modernized docks still handle some 3 million tons of cargo per annum, for Europe and the Baltic. But the buildings here have nevertheless retained some of their dusty Georgian charm.

Both the town and port were constructed by the firm of Sir Edward Banks, which, under the supervision of John Rennie, also built Waterloo, Southwark and London Bridges. Sir Edward was once commemorated by the Banks Arms Hotel, the first building to be completed here. After Banks's death it was renamed the **Lowther Hotel**, for Sir William Lowther, the first chairman of the Aire and Calder Company. While most of the workers' cottages have gone, many original buildings still survive, including the church of **St John the Evangelist**, which the company helped to build. Its spire is a well-known landmark. Twin water towers, old and new, known as the **'Salt and Pepper Pots'**, are the others.

On the river itself is the **Ocean Lock** (1938), which was used by coastal ships, while smaller craft passed through the **Victoria Lock** (1888) and **Ouse Lock** (1838). The latter is no longer in use. Both smaller locks are Grade II listed, as are several other buildings in the town. (Nothing, sad to say, gets the accolade of Grade I.)

There are eight docks at

Goole (a ninth was filled in). They include **Barge Dock** and **Ship Dock**, both built in 1826, **Ouse Dock** (formerly the 'Steam Ship Dock'), which dates from 1836, and **Railway Dock**, built in 1848 by the Aire and Calder Company as an interchange with the London and Yorkshire Railway.

Running parallel with the southern edge of the docks is the **Dutch River**, named in honour of the great engineer Cornelius Vermuyden, who canalized part of the River Don in the early seventeenth century. At **South Dock**, built in 1910, is the restored **No. 5 Hoist**, designed by W. H. Bartholomew – the inventor of the 'Tom Puddings' – which lifted the containers out of the barges and tipped the coal into waiting ships. (Bartholomew, who lived from 1831 to 1919, spent an extraordinary sixty-six years working for the Aire and Calder in one capacity or another.) Past the dry dock, a small basin to the north is where Humber keels removed their masts, and other sea-going tackle, before proceeding up towards Leeds. The **Yorkshire Waterways Museum** is just past the basin.

At the **Dutch Narrows Lock**, just before the **Doncaster Line Bridge** carries the railway over the canal, the Aire and Calder Navigation begins. This section, engineered by John Rennie and opened in 1826, is officially the **Knottingley and Goole Canal**. From here to Newbridge we pass by flat farmland, interrupted only by the small village of **Rawcliffe Bridge**. Just beyond, on the north side, are the **Sugar Mill Ponds**. They are the first of many examples we will encounter of recent efforts to transform the scars left by now-abandoned industry. In this case, a glucose factory (hence the misleadingly attractive name), whose closure had rendered the ponds an eyesore. Under the 'Changing Places' programme, they are being refashioned into a nature reserve – a process in train on other parts of the Aire and Calder.

At **Newbridge** the canal takes a 90-degree turn south, still running parallel with the Dutch River, down to **Beevers Rack**. Here the Dutch River continues its journey southward, to meet the waterway network of Sheffield and South Yorkshire. The Knottingley and Goole Canal heads west, past **Southbridge Reservoir**, now much in use for sailing. From here the **New Junction Canal**, finished in 1905, takes a more direct route than does the Dutch River, to Doncaster and on to Sheffield.

With the canal now accompanied by the doughty New Fleet Drain South, the flat arable landscape continues as we pass through the village of **Pollington**, where an attractive example of a lock-keeper's cottage stills stands, as do several Georgian farmhouses, including **Pollington Hall**. After Pollington we make a brief foray into North Yorkshire, now in the company of the New Fleet Drain North, past more arable land, dotted with an occasional wood. The canal passes by **Heck** (inevitably) and on to **Whitley Lock**, in sight and hearing of the M62.

From here towards Knottingley the canal is reinforced with steel piles, to counteract the subsidence caused by mining. We travel through more flat farmland, with woods in the distance, to the giant **Kellingley Colliery**, on the north bank. One of the newest deep mines in Britain – begun in the late 1950s – it is due

to close. Here 'Tom Puddings' can still be seen moored along the bank, bereft of cargo since the coal now goes by road.

At **Bank Dole Lock** William Jessop's **Selby Canal**, completed in 1778, runs six miles up to the Yorkshire Ouse. This canal – and with it Selby itself, previously an important inland port – were rendered far less useful, though not redundant, by the new Knottingley and Goole Canal in 1826.

At **Knottingley** we enter the West Riding of Yorkshire. The town was an inland port in the seventeenth century, once the upper limit of the navigable River Aire, until it was superseded when the original Aire and Calder Navigation was completed in 1699. It remained an important boat-building centre. One of its most famous boatyards, John Harker's, still survives, nowadays as a maintenance rather than boat-building business. It's on the south bank, just after Bank Dole Lock. Knottingley became known for its pottery and glass-making in the 1870s. The pottery business died out in the 1940s, while the glass manufacture continues. The town was an important source of limestone, too, and along the side of the canal a former quarry has been turned into a mini nature conservancy area.

Knottingley is umbilically linked to **Ferrybridge**, which is actually where the Knottingley and Goole Canal comes to an end, it having been decided to extend the line early in the planning stage. Here, past the **Ferrybridge Flood Lock**, where the canal rejoins the Aire, are the eight elephantine pot-like chimneys of the **Ferrybridge 'C' Power Station**, built from 1962. It once hoovered up vast amounts of coal brought in from thirty collieries, a quarter of which came via the canal. The traffic came to an end in 2002.

After passing under a handsome eighteenth-century bridge, which once carried the Great North Road (and ignoring its ugly modern successor), we continue on to Castleford, on what's known as the **Five Mile Pond**. It's surrounded to the north by 'ings', a Viking term for water meadows. The subsidence caused by opencast mining has caused much water-logging here, and the area has recently been converted into an RSPB wetland centre. Some spoil heaps are still in use – the ones that are disused have been planted with trees. The same process – areas ravaged by industry being given a new lease of life as nature reserves – continues right the way up to Hunslet, in Leeds.

From Fairburn the canal is surrounded by the sad remnants of former collieries that once thrived along its banks. The **Old Wheldale Colliery Loading Basin** survives only in a rusting coal-loading chute that sadly awaits a shipment that will never come again. Because of the flooding caused by subsidence, the lock-keeper's quarters at **Bulholme Lock** are built on stilts.

We then arrive in **Castleford**, which in the nineteenth century became a wealthy town thanks to the collieries on its doorstep. At **Hargreaves Boatyard** in Castleford the carrying company based here shipped over a million tons of coal a year to the power station at Ferrybridge 'C', until the trade tailed off in recent years. **Queen's Mill**, formerly Allinson's, the world's largest stone-grinding flour mill, is now a heritage centre. Its wall still proudly proclaims, in paint, 'Allinsons Stoneground Flour'. It's situated on the River

Aire, while the canal cuts through a loop of the river.

At **Castleford Junction** the Aire meets the Calder Navigation, on its way to Wakefield. The confluence led to the local rhyme,

That's why the Castleford girls are so fair,
They bathe in the Calder and dry in the
Aire.

The Aire, once again canalized, and indeed recently diverted (and enlarged) because of subsidence, now travels past pleasant countryside to **Allerton Bywater**. Here the colliery of the same name was the last deep mine to operate in the area, and an important customer for the services of the 'Tom Puddings', which took the coal down to the port at Goole. Former collieries at **St Aidan's**, north of the canal, have been converted into one of the largest wetland sites in the country, due to be managed by the RSPB. Just before **Lemonroyd Lock**, the canal and river part company again, not to rejoin each other until Knostrop Flood Lock, in the centre of the city.

The Leeds city border is passed between **Woodlesford** and **Fishpond Locks**.

In an otherwise bleak area surrounding the canal, scarred by open-cast mining, there is to be found the most northerly vineyard in England, as well as **Temple Newsham Country Park**. **Temple Newsham House** was once owned by Cardinal Wolsey (hence, perhaps, its being known as the 'Hampton Court of the North'), and was remodelled up until the reign of Charles I, before being restored again in recent times. In 1922 the estate was sold to Leeds Corporation, with covenants insisting it remain a public park. There's also a **Home Farm**, with many rare breeds.

To the south **Rothwell Country Park** has a very different set of antecedents, once being the home of Rothwell Colliery. When it closed in 1983 the area became a wasteland. It too was restored, and opened in 2000. As we reach the eastern suburbs of Leeds, at Stourton, we pass by **Thwaite Mills**, now an industrial heritage museum. Established as a fulling mill in the late seventeenth century, owned by one of the many millers who opposed the original navigation, it later turned its hand to pottery. The present building dates from 1823. The relic of an old steam derrick stands by the canal.

Passing through the urban suburb of **Hunslet**, we reach **Knostrop Fall Lock**. Here are the remains of a gigantic swing bridge built for the railway line that once crossed the river – but was never used. Past **Knostrop Flood Lock** – which is a stop-lock, rather than a pound-lock – we rejoin the River Aire. Here on the south bank of the canal are the derelict, but spookily impressive **Hunslet Flax Mills**. Built in the 1840s, staffed at their height by 1,500 female workers, they were abandoned in the 1970s and are still awaiting their fate.

Further along, on the opposite side of the canal, are the even older **Bank Mills**, another flax mill, established in the 1790s and in operation right up until the 1980s, when oil barges still made deliveries. Just past the mills the river and canal split for a short distance, the Aire making a loop to the north.

As we near **Leeds Lock**, attractive modern developments rub shoulders with historic buildings, adapted for new use. The splendid **Royal Armouries Museum**, built in 1996, houses 75,000 exhibits of arms and armour. The museum is part of the redeveloped **Clarence Dock**, which, because of the great numbers of spuds that were once unloaded here, became known to boaters as 'Tattie Dock'.

The canal is rejoined by the Aire opposite **Brewery Wharf**, another new development, based on the site of Tetley's Brewery, which was founded in Hunslet in 1822. The famous Art Deco Tetley's building was finished in 1931. The brewery was closed in 2011. It's now an art gallery, **The Tetley**.

The Aire now passes under the elegant **Leeds Bridge**, with its fine, shallow arch, designed by John Rennie. Nearby, in the city centre, the imposing High Victorian **Corn Exchange** a nd **Town Hall** are testaments to the great wealth created by England's third-largest city's importance as a wool and textile centre – and the navigation's role in making this expansion possible. Just to the west of the bridge the Aire and Calder Navigation becomes the **Leeds and Liverpool Canal**.

The Navvies

Many are the descriptions that have come down to us of the great celebrations attending the opening of canals, with cannons being fired, bunting hung out, oxen roasted, toasts drunk – and drunk again – valiant addresses made and backs soundly slapped. We know the names of the subscribers, and management committees. We have a few tantalizing glimpses of life on the canal from some of the rare journeys undertaken by curious observers. Of the men who literally built the canals we know virtually nothing. And most of what we do know is bad.

They were originally called 'cannallers' or 'bankers'. The term 'navigator' gradually came into use towards the end of the century. (The word 'navvy' was not used until 1832.) The best guess seems to be that rural labourers were hired to do the work, possibly aided, in the case of the Bridgewater Canal, by the duke's own mine-workers. But they could hardly have been spared for long periods. The use of farm labourers was a bone of contention with local landowners, who complained that their workforces were being tempted away by the higher wages canal work could offer. (They got their own back during harvest, when they could outprice the canals and get the farmworkers back. Joined perhaps, by other navvies.) Later, 'bankers' were recruited from Ireland and Scotland.

By 1800 there were an estimated 50,000 navvies at work – and at play.

Although most of the labour seems to have been local, as canal building developed gangs of navigators began to follow the work, living in riotous assemblies near the canals, causing much grief to the local populace with their hard-drinking and violent ways.

At one village pub they had an argument with a baker from whom they had bought their bread. The dispute escalated to the point where the navvies chased the innkeeper away, drank his beer and stole his pub sign. Then they caught up with the unfortunate baker and pelted him with his own loaves. They were heading in the direction of the next village when the local yeomanry arrived to restore order.

Canal boat propellers, 1894

The bad reputation of the canal navigators would continue down the years:

> *It is the general custom to employ gangs of hands who travel from one work to the other, and do nothing else. These banditti, known in some parts of England as 'Navies,' or 'Navigators,' and in others 'Bankers,' are generally the terror of the surrounding country, they are as completely a class by themselves as the Gipsies. Possessed of all the daring recklessness of the Smuggler, without any of his redeeming qualities, their ferocious behaviour can only be equalled by the brutality of their language. It may be truly said, their hand is against every man, and before they have been long located, every man's hand is against them; and woe befall any woman, with the slightest share of modesty, whose ears they can assail . . . Crimes of the most atrocious character are common, and robbery, without any attempt at concealment, has been an everyday occurrence, wherever they have been congregated in large numbers.*

(The irony of this description, written in 1839, is that the author was comparing the shocking conduct of the canal navigators with the exemplary behaviour of the navvies then building the railways, at that time local men. He was, of course, a railway lobbyist. Only for a short time would railway navvies be held up as paragons of considerate construction.)

The canal-builders could, had they been able to read such calumnies, have congratulated themselves that their own reputation would never be half as bad as that of some of the men working the boats.

The Leeds and Liverpool Canal

Introduction by John Sergeant

Liverpool makes a fitting climax to this canal journey from Leeds. I started my career as a journalist working here as a local reporter in the 1960s and I don't need to be told how Liverpool has been mightily improved. It seemed dark and dismal, with its black buildings and grim signs of long-term economic decline. Now it has regained much of the swagger and style of a successful city.

In the summer, around the rejuvenated Albert Docks, it has the air of a permanent festival, thronged with visitors and an array of old sailing ships. In the midst of it all there are a dozen or so moorings for privately owned narrow boats, the present-day successors to the generations who plied their trade along the longest canal in Britain – the one from Leeds to Liverpool.

We made the final part of our journey from the outskirts of the city on a charity barge. On board were a lively group from Liverpool's Women's Institute who had spent much of the summer baking cakes to raise money to send a choir abroad. The skipper took us along a new section of canal, allowing us to travel past the Liver Building and the famous waterfront. It had cost £26 million pounds, but a quick check with the WI confirmed that the money had been well spent. On a lovely, hot day, they were bursting with pride at their city, reborn.

When we first climbed on to a narrow boat in the centre of Leeds one of the main themes emerged of this particular canal journey. The route takes you through some of the great mill towns of the north. Wool and cotton were imported in bulk to keep the spinning machines busy. Many thousands of people were employed to spin and weave, with their finished goods in demand across the world. Leeds, which had been a modest wool town, became the third-largest city in the country.

For decades this canal, once a vital artery from east to west across the Pennines, was allowed to sink into decline. It was regarded as an embarrassing problem which it seemed could only be resolved by large amounts of taxpayers' money, increasingly difficult to justify. But not only in Leeds and Liverpool, right along the route, there were signs of dramatic renewal. Instead of turning their backs on the derelict remains of their industrial past, people look to the canal to enjoy the modern amenities of the waterway. New offices, homes and restaurants are spreading along the banks.

Much of the restoration has been helped by public funds and government assistance. One of the most impressive sights is the restored mill founded by the famous industrialist Sir Titus Salt, now a world heritage site. Saltaire was a company town. Employees could be born there and work there all their lives. One of those, now in his nineties, told me how hard life could be. He remembered seeing men in the 1930s working all day with shovels and wheelbarrows emptying tons of coal from the barges.

It would be easy to describe the Leeds and Liverpool Canal as a series of encouraging vignettes, a string of improvements. But there are stretches of the canal which are sadly run down, where the waterway attracts rubbish and boarded-up buildings cry out for redevelopment. The canal enthusiasts I met are in no doubt that much still needs to be done. The first-time visitor should not expect to be thrilled by every stretch of the canal. But this journey contains some of the finest countryside in Britain, with some breathtaking views, and it gives those with a love of history the chance to contemplate the extraordinary changes which have shaped people's lives over the past two centuries.

At 127 miles long, the Leeds and Liverpool Canal is the longest British canal once in the hands of one company. But in its earliest years it didn't always look that way.

If to take a journey through the canals and navigations of the West Riding of Yorkshire is to embark on a tour of Rugby League, so a journey through the Lancashire part of the Leeds and Liverpool Canal is to encounter the world of Association Football, as the oval ball gives way to the round. The homes of historic football clubs like Burnley, Accrington and Blackburn Rovers line the canal, while a line branches off to find Preston North End – all to become founders of the Football League in 1888. (Of course some places had a foot in both camps.)

The history of the Leeds and Liverpool is also a game of two halves. The textile industries on each side of the Pennines developed quite different traditions. The ultimate success of the Leeds and Liverpool Canal came from the fact that it was able to serve both counties, with their competing demands and expectations. It played to its weakness, as it first appeared.

The West Riding of Yorkshire looked east, shipping its goods by navigable rivers, such as the Aire and Calder, to Hull and then down to London, and markets in Europe. Lancashire looked in the opposite direction, with links to the once-important Irish linen trade (eventually suppressed), Irish dairy and farming produce and the Americas. Lancashire's rivers were fast-flowing and generally unnavigable. Instead, they were tapped for power, and mills sprang up along their banks. In Liverpool and its surrounding area were sugar refining and salt and iron works. All needed coal. And the surrounding roads were poor.

A navigation in Lancashire did, after much travail, get made. This was the Douglas Navigation, linking Wigan – the main producer of coal in the region – to the Ribble estuary to the north. First proposed in 1720, temporarily abandoned due to embezzlement and only finished in 1742, it was to play an important, if fractious, part in the Leeds and Liverpool story.

The impetus for a trans-Pennine canal came from Yorkshire, where the wool merchants of Leeds, Wakefield and Halifax – and especially those of Bradford, who had no access to a navigation at all – wanted to access Liverpool, and thus the American trade. At this point, merchandise travelling across the Pennine passes had to be taken by trains of packhorses, carrying their goods in panniers. A slow, inefficient and cumbersome means of transport. Meanwhile, a proposal to make the Aire navigable from Bingley to Skipton, between which were rich sources of coal and limestone, was successfully opposed by landowners.

In July 1766 a public meeting was held at the Sun Inn, Bradford, called by John Longbotham, a Halifax engineer, and a Quaker merchant from the same town, the encouragingly named John Hustler. Subscriptions were raised, and Longbotham was engaged to survey the route. His plan was for the canal to start at Leeds (where, of course, it would link up to the Aire and Calder) and travel up through Bingley and Skipton, by the side of the upper Aire. A branch would also be built to Bradford. He would then build a link from the Aire to the Ribble, via a natural gap in the Pennines. The canal would end at Preston – then an important inland port. From here one branch was to go to Liverpool, another to Lancaster, which at that time rivalled Liverpool as a port.

During the course of surveying the branches, Longbotham seems to have taken a leaf out of James Brindley's book and now came up with a bolder scheme: to build a further section of canal from Leeds straight down to Hull (the Aire and Calder would have strongly resisted that). He was persuaded to tackle just the Leeds to Liverpool line. The route surveyed, Brindley himself, with his trusted assistant Robert Whitworth, was asked to check it. Longbotham, significantly, favoured a broad canal (and was worried that Brindley wouldn't).

The Bradford promoters now began to try to sell the idea in Lancashire, but only two years later was the first public meeting held there. This lukewarm response to Longbotham's canal was a sign of things to come. Nevertheless, after a subsequent joint meeting of supporters from the two counties, a Bill was presented to parliament. Submitted too late to be considered for the 1769 session, it had to be withdrawn.

This gave a chance to the Lancashire merchants to put a spoke in the wheel. It was clear from Longbotham's route that the primary purpose of the canal was to get Yorkshire woollens to the Irish Sea as quickly as possible

– whether by Liverpool, Preston or Lancaster. This hardly fired the blood on the west side of the Pennines. What the canal defiantly didn't do was pass by Wigan, the major coal centre of this part of Lancashire, nor any of the major mill towns. There now ensued an ongoing squabble which dragged on for more than twenty years.

Woollen mills next to the Leeds and Liverpool
Canal at Saltaire, 1860

The Liverpool group, to the dismay of the Yorkshire merchants, took matters into their own hands and paid for their own survey of the Lancashire section. Subsequently a bewildering variety of different routes through Lancashire was suggested, most of them involving the Douglas Navigation, and thus Wigan, as well as the mills. The Yorkshire interests were aghast. What they desperately wanted to avoid was a canal that made its way slowly around the mill district, calling in at any towns that took its fancy. The Lancashire faction insisted that it do just this. The red and white roses were at an impasse, not least because Longbotham's route to Preston was a contour line, passing through relatively flat terrain. If the canal was diverted through the areas the Lancashire promoters favoured, not only would it be 17 miles longer, it would need deep cuttings and more locks, making it far more expensive to build, and navigation slower still.

Some of the Lancashire supporters dropped out. Nevertheless the Yorkshire promoters, who held the majority of the shares, won the day – for now. A compromise route, which had been suggested by Longbotham, would be the one proposed to parliament. It would run from Colne to Whalley (with a branch to Burnley), across to Newburgh, on the Douglas Navigation, and thence to Liverpool. This line would miss out the important mill towns of Blackburn and Chorley entirely.

And that, the Lancashire interest thought, would be a pity. They turned out to be absolutely right. The canal was to play a vital role in the court of King Cotton.

Cotton

Cotton was, unlike the long-established woollen industry, fairly late to make its mark in Britain. The manufacture of pure cotton had only become well established by the 1750s. By the end of the century it had become Britain's most important industry. Cotton was a truly global trade. Through the port of Liverpool huge quantities of raw cotton were imported from the slave plantations of the West

Indies and America. The finished goods, until the 1860s and beyond, made up around half of Britain's total exports.

Cotton drew on several advantages. The government was willing to give it aggressive backing, for one thing. The import of Indian calico had been, at the request of the British linen industry, banned in 1700. This, ironically, gave homegrown cotton a clear run in the domestic market, to the discomfiture of other cloth-makers. Later, through the offices of the East India Company, the Indian cotton industry – which had led the world – was suppressed altogether. India became one of the principal markets for finished cotton goods from Lancashire.

As in effect a new industry, cotton was far more willing to welcome innovation than were the more traditional makers of textiles. Elsewhere, not only were spinners and weavers opposed to machines that would take away their jobs; the clothiers too saw little reason to change a winning formula.

Cotton could start more or less from scratch. So it was this industry more than any other that benefitted from a series of dynamic inventions that were to revolutionize the making of textiles, fire the industrial revolution and create, more than any single factor, the factory system, in all its glory, and all its ugliness.

In 1733 John Kay had invented the flying shuttle, which doubled the pace of weaving. Hand-loom weavers didn't thank him for his efforts to put them out of business, and he was physically attacked wherever he tried to set up shop. Nor could he collect on the royalties due on his patents, eventually moving to France to try his luck with the government there.

Richard Arkwright
(1733–92),
engineer

Although weaving had been accelerated, the problem was that the spinners couldn't keep up the supply of yarn. In 1738 Lewis Paul, still a somewhat obscure figure, invented a method of spinning that used rollers. For reasons that are unclear, it didn't really take off until fifty years later.

Meanwhile a most important breakthrough came with the invention of the 'Spinning Jenny', by James Hargreaves, in 1764. The jenny could spin six threads simultaneously, eventually enabling up to eighty to be worked at the same time.

The Spinning Jenny was followed by a significant advance – a frame for spinning that was powered by water. The credit for this was seized by the man who made a fortune from it, the first cotton magnate, Richard Arkwright. It worked on a similar principle to the frame developed by Lewis Paul but was far more powerful.

Arkwright almost certainly didn't invent it, and in 1785 his patent was taken away when he proved unable to give a satisfactory account of its genesis.

Arkwright, though he may have been no inventor, was, like Josiah Wedgwood, a spectacularly successful business-man, and a visionary. Arkwright's Cromford Mill, built in 1771 on the banks of the fast-flowing River Derwent in Derbyshire, was the first cotton spinning mill to rely on water power. (The complex still survives, as the Derwent Valley Mills World Heritage Site.) Like Wedgwood again, Arkwright was an early champion of canals, and the Crom-ford Canal was built partly to serve his mills.

The final crucial breakthrough in spinning came in 1779, when Samuel Crompton invented a machine that became known as the 'mule', so-called because it was a cross-breed between the Spinning Jenny and the water frame. Crompton unwisely took out no patent and thus, in a striking mirror image to the case of Richard Arkwright, made no money whatever from his ingenious contraption.

An early, though doomed, attempt was made to make weaving more mechanized thanks to the invention of an Oxford don, Edmund Cartwright, who developed the 'line shaft' in 1784. The shaft connected to the power source and then, through a system of wheels and pulleys, drove several frames at once. Cartwright left academe to put his ideas into practice, but his efforts didn't please hand-loom weavers, and his factories were attacked, with 400 looms being burned.

With the line shaft out of the picture, the abundance of yarn being produced put skilled weaving at a premium. The weavers of Bolton went so far as to wear five-pound notes ostentatiously tucked into their hat bands. But this 'loadsamoney' era couldn't last. Steam had begun to take over from water power in textile manufacture in the 1780s. As the inexorable tide of mechanization swamped the textile trade after the end of the Napoleonic Wars in 1815, weavers ended up on the breadline, as more and more power looms put them out of business.

Though it was made at Cromford and elsewhere, cotton was ideally suited to Lancashire. The soft Pennine water, the dampish climate, the fast-flowing rivers and the port of Liverpool, through which raw cotton was imported, made a perfect fit.

Arkwright opened the world's first steam-driven textile mill in Manchester in 1781. The town, from pretty much a standing start, became the largest centre of cotton making on the planet, known by Victorian times as 'Cottonopolis'. It was a vital component in Britain's trade. It was less successful as a centre of gracious living, as factories (in which workers lived, to begin with), then tenements, then subdivided tenements, which became insanitary and festering slums, grew up around the new mills.

Manchester already had its own canal, thanks to the Duke of Bridgewater. The Lancashire interests were determined that the smaller cotton mill centres – of Nelson, Burnley, Accrington, Blackburn and Chorley – would have theirs too. The Manchester pattern was to be repeated, in miniature.

After a slow start, the woollen industry of the West Riding was also mechanized, allowing it to eclipse its rivals in East Anglia and the West Country. It could get its hands on huge quantities of coal, thanks to the canals. They couldn't.

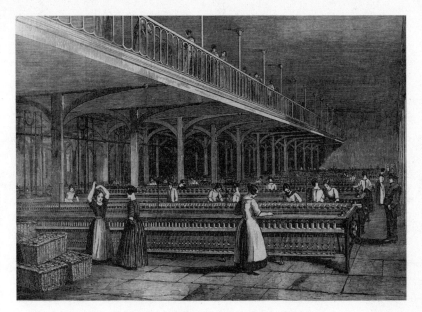
Women working in a cotton mill

The Leeds and Liverpool was authorized in 1770, and work began, starting at each end simultaneously. James Brindley was offered the post of engineer, but – for once – sensibly declined on the grounds that he was too busy. John Longbotham was appointed as chief engineer instead.

But if the Yorkshire party now thought it was game over, they were wrong.

At first, things went well. Negotiations were made to buy the Douglas Navigation (which succeeded in 1772). The line from Liverpool to Newburgh (on the Douglas Navigation) was opened in 1773, as was that from Bingley to Skipton, where, as was so often the case when canals got to work, the price of coal was halved.

At Bingley, Longbotham engineered one of the great

achievements of British canals. The Bingley Five-Rise staircase of – broad – locks. They were completed in 1774.

Bingley Five-Rise locks around 1900

Now the canal fell foul of the American War of Independence. Not only did the war restrict the supply of raw cotton, it also squeezed the supply of capital, and work on the remaining sections of the canal came to a standstill. Not until 1790 was further building done to complete the middle section. As before, a hiatus gave the Lancashire cotton interests the chance to wrest control of their part of the canal back into their own hands.

Longbotham having retired, Robert Whitworth now took over as engineer. Having surveyed the line, he suggested an alteration that pleased the Lancashire cotton owners. Diverting the line through Chorley, he decided, would save on a large aqueduct, originally planned for Whalley. Whitworth also took advantage of the advances in canal engineering that had been made during the previous couple of decades and suggested lowering the summit, which necessitated a 1,640-yard tunnel to be cut at Foulridge. Reservoirs would need to be constructed nearby.

Promoters of other canals now got in on the act, trying to seduce the Leeds and Liverpool into dalliances with them. The Lancaster Canal was on the starting blocks, and so too was the Manchester, Bolton and Bury Canal (which earned the wrath of the Duke of Bridgewater, who didn't want a canal taking Wigan's coal into Manchester to compete with his).

In order to explore these possibilities, Whitworth surveyed a new route, which went through Burnley, Accrington, Blackburn and Chorley – exactly what the Lancashire proprietors had always been asking for. Because by now the area was even more populous than it had been twenty years earlier, the Yorkshire partners finally agreed to this route being adopted, giving the line of the canal we know today. King Cotton had prevailed.

Another reason for the change of heart was that by this stage the Rochdale Canal had already been suggested. As its proposed route, from Halifax to Manchester, and thence on to Liverpool by way of the Bridgewater Canal, was far shorter than that of the Leeds and Liverpool, it was decided to take full advantage of the traffic in the cotton district. An Act authorizing the final route was passed in 1793.

The notion of joining up with the Manchester, Bolton and Bury Canal came to nothing – though the Lancaster Canal did play a walk-on role in the Leeds and Liverpool Canal. Ten miles of its southern section became a part of the main line, and it forms a noticeable dog-leg, as the canal heads south to Wigan and then sharply north-west.

Work did not progress quickly. There were the usual problems with landowners, the usual problems with contractors and the usual problems of things turning out to be harder than had been envisaged. The section from the summit down to Wigan was a particular bugbear.

There were many challenges. To traverse the valleys giant embankments had to be built, the biggest and most admired of which being the gargantuan 60-foot embankment at Burnley. The summit tunnel at Foulridge had to be dug through rotten rock and took six years (and much money) to construct. Meanwhile a seven-arch aqueduct was built over the River Aire at Dowley Gap.

Money was increasingly tight. The company was in debt, and trying to decide whether to spend its income in paying down debt, pressing ahead with construction or paying dividends. Only in 1816 was the whole canal open for business.

Although it was the first of the three trans-Pennine canals authorized, the Leeds and Liverpool Canal was actually the last to be completed. The Rochdale Canal had been finished in 1804 and became a serious rival. The Huddersfield Narrow Canal opened in 1811. Despite serving important mill towns, such as Hebden Bridge (famous for its corduroy), Rochdale and Oldham, it was unsuccessful as the sheer number of locks, and the enormously long Standedge Tunnel, made traffic slow.

The Leeds and Liverpool was a broad canal, meaning that Humber and Yorkshire keels could use it. But it was also, rather bizarrely, a short canal, in terms of its locks. The characteristic Leeds and Liverpool 'short boats' were only 62 feet long, and the locks – as always – were built to accommodate one boat of the size anticipated. This meant that narrow boats, at around 71 feet in length, could not use the canal, with the exception of the section between Wigan and Liverpool, in which the locks were 72 feet long. (They were designed, not with narrow boats in mind, but for the use of the Mersey 'flats' which had long used the Douglas Navigation.)

Barges carrying raw cotton to the mills

But in fact the through route wasn't the reason why the Leeds and Liverpool was the most successful of the three Pennine canals. As in all canals, the average journey was around 20 miles. In any case, the perennial problems with water supply at the summit would have restricted any great volume of through trade.

Instead, each county used its section of the waterway predominantly for its own local traffic. After all the agonizing on the part of the Yorkshire wool merchants, the need to transport their goods right through to Liverpool was not, as it turned out, the canal's primary function. As in nearly all canals, coal, by far, provided the majority of tonnage carried.

It would be a mistake, however, to assume that the canals at this period only carried the big-ticket cargoes of coal, ironstone and limestone. A newspaper report from Blackburn in 1810 gives us a rare insight into the general merchandise that arrived – in small volumes – by canal. It also gives us a hint as to how they transformed not just local industry but also local consumption, habits and tastes. Goods arriving included flour, beans, raisins, molasses, nuts, malt, porter, hops, soap, yarn, tow, woollen cloth, flax, oil, tallow, paint, pots, cast-iron pillars, gunpowder and snuff.

Further sections of the canal were built after the main line was completed. The seven-mile Leigh Branch, first proposed in 1800, allowed it to join the east end of the Bridgewater Canal in 1820 – after the usual prevarications and frustrations. It too was built only for boats 62 feet long – but was later modified to accept vessels of 70 feet. This gave the Leeds and Liverpool ready access to Manchester. In 1846 – having been prevented from doing so by the proprietors of the Liverpool Docks – the canal finally reached the Mersey itself, with the building of the Stanley Dock Branch.

The Leeds and Liverpool Canal
– the journey itself

Burnley Locks

With its attractive recent water-side developments at the start and end of the journey, and its spectacular scenery either side of the Pennines, the 127 miles of the Leeds and Liverpool Canal have riches in abundance. As Hugh McKnight put it, in the *Shell Book of Inland Waterways*, 'For the sustained magnificence of its scenery through the moorlands of York-shire and Lancashire, the Leeds and Liverpool can justifiably claim to be first choice among cruising enthusiasts.'

Time was, and in recent memory, when there were also dramatic scenes of the industry that made the revolution, espe-cially in the cotton mill towns of

Lancashire, which resembled an L. S. Lowry painting brought to life. (And with them, of course, the slums, and hopeless poverty written about by George Orwell.) These towns now are but a shadow of their former selves, their industrial vigour drained away. But, as always, the canal stretches green fingers into the most intractable landscape at every point along the way. And many old canalside mill buildings and warehouses do survive, some put to a variety of new uses.

The vernacular construction material – millstone grit, or gritstone, a hard sandstone that weathers almost black – gives a distinctive, rough-hewn appearance to many of the buildings that line the canal, some of which could have come from a Brontë novel.

There are some notable eccentricities. As built, some locks had no ladders on the lockside; boatmen climbed up via iron rungs set into the wooden gates. Strange 'jack cloughs', which had to be lifted up and down, rather than turned, operated some of the paddles – an ingenious but not entirely effec-

tual variant. Most have been replaced. The maximum boat length is 62 feet for a through journey, breadth 14 feet 4 inches.

We begin our journey in the heart of **Leeds**. The **River Lock** is where the Leeds and Liverpool formally joins the Aire and Calder Navigation. Around here the waterside in the heart of the city has been smartly redeveloped. One nearby example is **Granary Wharf**, a shopping and residential complex built into the brick tunnels – or 'Dark Arches' – under the former main railway station, where coal was once shipped to a local power plant from the swirling River Aire. At the next lock along was the HQ of the Leeds and Liverpool Company, **Office Lock**.

As the canal heads west, through this mixture of old and new buildings, including the redeveloped **Castleton Mill** and **Tower Works**, we come to **Armley Mills Industrial Museum**. The mill itself dates from the sixteenth century. By 1788, it was the world's largest woollen mill. It was rebuilt in 1805 after a fire. We then come to **Kirkstall**, with its imposing ruined Cistercian abbey, founded in 1152. Here the old **Mackeson Brewery** has been turned into a hall of residence, another good example of an old canalside building put to new use. Near **Kirkstall Lock** is the site of the historic forge founded by the monks.

At **Newlay** we begin to leave the Leeds suburbs and pass through a wooded area that leads to **Rodley Nature Reserve**. Rodley was once a busy canal centre, as the **Rodley Barge** pub reminds us. From here the scenery is mostly rural until we reach **Shipley**, once known for its woollen mills. Today it forms part of a large conurbation with Bradford. At **Junction Mills** the canal used to be joined by the short Bradford Canal until it was closed in 1866, as a public health hazard, owing to its extreme pollution – even by mid-Victorian standards.

Adjoining Shipley on the western side is the famous model village of **Saltaire**. It was founded by a self-made entrepreneur and philanthropist, Sir Titus Salt, whose fortune had been built on alpaca, which he turned into 'an affordable luxury cloth'. In 1834 he'd found a shipment of the wool lying around in Liverpool Docks, unsold. Spotting its potential, he took away a sample, made some experiments and then came back and snapped up the lot. His rise to prominence coincided with the increasing mechanization of the woollen industry in Yorkshire. In 1851 he founded Saltaire, with the twin purpose of allowing his workers to escape the slums of Bradford and to make use of the canal – as well as the railway.

Two years later the village had been built. **Salts Mill**, six storeys high and over 500 feet long, with its splendid gold-coloured Italianate façade, was then the largest industrial building in the world. A second mill towers over the canal on the other side, and other original buildings still stand. By 1871 there were 820 homes here. Statues of lions – rejected as too small for Trafalgar Square – found a position too, guarding the school. Unlike Richard Arkwright a century before, Salt – a strong Non-conformist – didn't build a pub for his workers. But neither was there a pawnshop. The park by the canal was described in 1903 as 'one of the most beautiful in the world'.

Now comes an attractive

section of the canal, as we head past woods, over the aqueduct at **Dowley Gap** and on to **Bingley**. Its Georgian Market Hall and Butter Cross bear witness to the fact that it was an important wool town. It prospered further thanks to the canal, which encouraged both mills and workers to migrate here.

After Bingley we begin to climb towards the Pennines thanks to a three-lock staircase, the **Bingley Three-Rise**. But this construction, raising the canal 30 feet, is merely the starter. Soon after comes the main course – the **Bingley Five-Rise** staircase, selected by Robert Aickman as one of the 'Seven Wonders of the Waterways'. This grandiose and highly attractive structure (unless, perhaps, you are operating the locks) climbs 60 feet, making it the steepest lock staircase in Britain. The upper pound is 17 miles long.

After the suburbs of **East Riddlesden**, near Keighley (it has a seventeenth-century manor house, **East Riddlesden Hall**), we pass by some original warehouse buildings on the canalside at **Stockbridge** and then into open country all the way to Skip-

ton, the 'Gateway to the Yorkshire Dales'. From here are fine views that will last all the way to the Pennines, with the **North Yorkshire Dales National Park** providing them ('dale' meaning valley).

After the little mill village of **Silsden** we pass through green wooded hills, over an aqueduct and through the pretty village of **Kildwick**, the border between the West Riding and North Yorkshire. At **Farnhill Wood** are sycamore, birch, alder and beech.

There are several swing bridges en route, such as the one at **Bradley**, with its attractive waterfront and Georgian high street. Here are superb views of the Aire Valley as well as the wilds of Ilkley Moor to the south-east.

Skipton is an attractive town, with its restored canal warehouse, **Holy Trinity Church**, founded in the twelfth century, and Norman castle lowering over the town (reached through the **Springs Branch Canal**). The ten miles from Skipton to Gargrave are some of the most beautiful on the canal, with lush Airedale scenery and remote moors. **Gar-**

grave is a pretty village, with a limestone Saxon church. The **Pennine Way** is nearby.

There are several turnover bridges along this stretch, as well as farmsteads. Just before **Bank Newton** the River Aire is crossed by the **Priest Holme Aqueduct**. Then we pass through the six **Bank Newton Locks**, and the former Bank Newton wharf. From here the canal takes a windy route through **East Marton**, with its curious double-arched road bridge, to the summit level. It starts at **Greenberfield Locks**, with their superb views of the Dales and distant mountains. Anthony Burton and Derek Pratt, in *The Anatomy of Canals – The Early Years*, thought: 'There are few, if any, more delightful lock flights than this. The scenery is inspiring, the views across the hills magnificent and the immediate surroundings of the canal fits perfectly into the wider picture.'

In nearby **Barnoldswick**, rather unexpectedly, is a Rolls-Royce engineering plant. Here the godfather of the modern canal movement, L. T. C. Rolt, was an apprentice.

Between this attractive village and another one, **Salterforth**, comes another stretch of fine, remote countryside, leading to **Foulridge Wharf**, with its old warehouses. We are now in Lancashire. **Lake Burwain** is nearby, used as a reservoir for the summit as well as for sailing. The canal then passes through **Foulridge Tunnel**.

The tunnel, like many others on the canal network, proved a handful. It was finished in 1796, having taken six years to build. Nearly a mile long, cut through rotten rock, it suffered from constant problems with water incursion, collapsing in 1824, 1843 and 1902. Its main claim to fame, however, is thanks to a tale of bovine endurance. In 1912, a cow called Buttercup fell into the water by the southern entrance. Rather than getting out while the going was good, the animal chose to swim the whole length of the tunnel. When she reached the other end, astonished observers revived her with brandy.

We now continue through pleasant farmland, to the seven **Barrowford Locks**, which lower us 70 feet from the summit. Here in the distance is beautiful mountainous scenery.

An aqueduct then takes us over the River Colne to **Nelson**, the first of the important mill towns that now line the meandering route through the heart of what was once industrial Lancashire. There are fine views of **Pendle Hill** to the west as we pass through **Brierfield** towards Burnley. Pendle Hill is indelibly linked to the persecution of – supposed – witches in England. In 1612 nine local women and two men were put on trial for witchcraft, the records of which provide one of the most detailed accounts of the mania for witchhunts during the reign of James I. Ten of the eleven accused were found guilty and hanged.

Passing by old mill buildings on the canalside, we now come to the astonishing **Burnley Embankment**. It is another of Aickman's 'Seven Wonders of the Waterways'. In *Navigable Waterways*, L. T. C. Rolt writes of it:

Three quarters of a mile long and over 60 ft high, this tremendous earthwork carries the canal in a majestic curve around and above the town of Burnley, presenting the canal traveler with one of the most striking industrial landscapes in Britain.

He was writing, however, in 1969. Since then, the heart has been ripped out of industrial Britain, and while it's true that some of the old mill buildings have survived both the death of the cotton industry and the planners, much of Burnley has lost its former glory. The embankment itself wasn't without its problems. Collieries to the east literally undermined it, causing subsidence.

After passing along the embankment, the canal turns sharply north-west and through the **Weaver's Triangle**. This was once the centre of the textile trade in Burnley, and some of the buildings have been well preserved. The **visitor centre** is based in an attractive former wharfmaster's house and old toll-office and celebrates the town's industrial heritage. Civic buildings such as the late Victorian town hall indicate the former prosperity.

Leaving Burnley, there are superb views across the **West Pennine Moors**. We then turn south, in a notable loop, to pass by the small town of **Clayton Le Moors** and the village of **Church**, where **St James's**

Church has a late medieval tower and stained glass windows designed by (William) Morris & Co. Here in Church the father of the future prime minister Sir Robert Peel had a factory making calico. The fact that the son of a plutocrat manufacturer, rather than of an aristocrat, became PM indicates a shift in the make-up of the Victorian ruling class. Several old canal buildings survive. We skirt round the outskirts of the former mill town of **Accrington** and then sharply north, and then west towards Blackburn. The M65, alas, is an almost constant companion on the whole section from Burnley to Blackburn.

We now pass through farmland on the way to **Rishton**, where calico was first manufactured in England on a large scale. And then we reach the suburbs of the important mill town of Blackburn. As with its great rival, Burnley, the decline of industry has done the town and its inhabitants few favours, and a once-thriving hub of the textile industry has suffered from the demise of its historic infrastructure and source of employment. However, as at Burnley, many

superb mill buildings do survive by the canalside, a heritage of its glory days when British cotton enveloped the world.

The impressive **Imperial Mill** still stands. It was built in 1901. Up to then, Blackburn had been principally a weaving town, and the mill was meant to prevent the need for yarn to come from outside. But, in reality, the great days were already over. Further along the canal, the **Daisyfield Mill**, the largest surviving corn and flour mill in the district, has been restored by the Canal and River Trust, and is now an office complex. The once-important **Eanam Wharf** has seen better days. One historic warehouse remains. Like the Imperial Mill, it is crying out for sensitive redevelopment. Thankfully, it has recently been designated as a conservation area, at least.

As we wind south-westwards through the Blackburn suburbs, over the River Darwen via an aqueduct, and through **Mill Hill** and **Cherry Tree**, the trees lining the canal create a pleasantly rural feel. After **Feniscowles** we leave the Blackburn conurbation, to make a dog-leg through a valley with attractive wood and pasture land, past an old paper mill, with its high chimney dominating the landscape. Then another aqueduct crosses the tiny River Roddleswell. At **Riley Green** the canal once more bears southwest, though an attractively wooded valley, past another former paper mill – which once produced bank notes – and a nature reserve at **Withnell Fold**, to the seven **Johnson's Hillock Locks** at **Wheelton**. Here timber bridges cross over the weirs, making this an attractive setting, with good views across the countryside. Up until 1932 the Lancaster Canal joined the Leeds and Liverpool here. The building of the M61 led to several sections being culverted, i.e. diverted into underground tunnels, to much protest. The Lancaster was a classic case of a Canal Mania project that never arrived. After its summit at Walton, a temporary tramroad was built to link it to Preston, which was meant to be replaced by an aqueduct over the River Ribble. The money ran out, and the aqueduct tramroad remained until its closure in 1879.

Our journey down to Chorley is now along the south end of the Lancaster Canal, suborned into the Leeds and Liverpool. This section of canal, nine miles long and lock-free, is known as the 'Lancaster Pool'. The navvies who built it are, reputedly, remembered in the name of the former mill area, now a shopping centre, **Botany Bay**. The reference is to transportation to Australia, the common fate of criminals at the time, which the locals perhaps hoped would be visited on the labourers who camped here, notorious for their wild-living ways.

The canal then skirts past **Chorley**, formerly another thriving mill town, down to **Adlington**, where an aqueduct at **Red House** takes us over the River Douglas and into the valley, with views of the West Pennine Moors to the east. The 1,200-foot **Pike Tower** can be seen in the distance, on **Winter Hill**. Now parallel with the river, the canal travels through more attractive woodland, past the beautiful gardens at the Victorian pseudo-Jacobean **Arley Hall** and through the **Haigh Hall Country Park**. Further down, the Lancaster Canal section of the Leeds and Liverpool comes to an abrupt full stop. The idea had been to try and join the Bridge-water Canal at Worsley, but this never came to fruition because of the objections of landowners. Instead, the Leeds and Liverpool proper takes matters back into its own hands and turns at right angles to descend 200 feet, in 21 locks, in 2 miles, to seek out the once-vital coal town of **Wigan**. (In the distance one could see the famous leaning spire of St Catherine's Church in Scholes, Greater Manchester. Rather sadly, it is being straightened.)

The start of the flight is deceptively rural, but the town soon comes into view. At the bottom of the locks the Leigh Branch of the Leeds and Lancaster Canal succeeded where the Lancaster Canal had failed and made a connection with the Bridgewater Canal at Worsley, in 1820. West of Wigan, the locks are longer, 72 feet as opposed to the 62 feet of those east to Leeds. Here at Wigan cargoes were sometimes trans-shipped on to the bigger boats.

Like the other industrial powerhouses of South Lancashire, Wigan is in the grip of post-industrial decline. But many attractive old industrial buildings are still to be seen on the canal-side.

The most infamous waterside location is, of course, **Wigan Pier**, which, in 1937, George Orwell chose as the symbol of the despair and hopelessness of the poverty, 'in the dreadful environs of Wigan', that went hand in hand with the great profits being made from industry. By the time he visited the town, the pier itself – a loading staithe for coal – had actually been demolished, although other buildings still survive. Orwell was left with the impression of 'a world from which vegetation had been banished; nothing existed except smoke, shale, ice, mud, ashes, and foul water'. In a bizarre twist, it was reconceived as a heritage centre in 1986, the 'Way We Were'. The 'Wigan Pier Experience' used Orwell's name and book as what it called 'a strong marketing tool'. It closed in 2007. 'The Orwell at Wigan Pier', a converted cotton warehouse, is a bar and restaurant.

After we leave Wigan we pass through woods and farmlands all the way to the outskirts of Liverpool, a few urban intrusions aside. We begin by following the course of the River Douglas, until it splits off at Parbold. There are many swing bridges along this section, along with a charming windmill at Parbold itself, built in 1794 and working until 1850. The Rufford Branch of the main line leaves just before Burscough, passing through the elegant **Junction Bridge** to meet the Douglas Navigation around Tarleton. Beginning with the **Ring O' Bells**, nearby, there are a large number of canalside pubs along this section.

The main line passes westwards through the former canal village of **Burscough**, with its old mills and cottages at the water's edge. At **Scarisbrick** marina we head south-west, then south, to Liverpool. **Scarisbrick Hall** was designed by E. W. Pugin, in the style of his father, A. W. N., one of the architects of the Houses of Parliament, and the great champion of the Victorian Gothic Revival. At **Halsall** in 1770 the first sod of the Leeds and Liverpool Canal was turned over.

From here to **Lydiate** attractive farmland follows, with more swing bridges, and more pubs, including what is thought to be the oldest in Lancashire, the **Scotch Piper**, at Lydiate, which dates from the 1320s.

After passing through the more urban **Maghull**, we travel through a generous contour of more open countryside and swing bridges, make a canal turn to pass north-east of **Aintree Racecourse**, enter the outskirts of Liverpool and travel along the side of **Rimrose Country Park**. **Litherland** and **Bootle** are thickly urban and industrial. Continuing south through **Bank Hall**, we pass several impressive old warehouses.

The canal then turns 90 degrees and heads for the Mersey, which the Corporation of Liverpool managed to prevent it joining until 1846, when the Stanley Dock Branch was built. Until then the terminus was at Pall Mall.

Stanley Dock and **Salisbury Dock**, which has an outlet into the Mersey estuary, were completed in 1848, and designed by Jesse Hartley. On the north side of Stanley Dock is the handsome **Tobacco Warehouse**, built in 1901, the largest brick warehouse in the world. It was abandoned and fell into disrepair in the 1980s. Plans are afoot to redevelop it.

The success of such dockside renovation is shown nearby. In 2009 the **Liverpool Canal Link** was opened, a brand new waterway running parallel to the Mersey and joining up Salisbury, Trafalgar, Waterloo, Princes and Canning Docks, passing right in front of the famous Liver Birds.

Working on 'the Cut'

River boatmen had long had a reputation for their bad language, violence and drunkenness, whether fairly earned or not. One female observer described the bargees on the Severn as little better than 'inland pirates'. The canal boatmen, at first at least, were cut of a different cloth. In 1771 a visitor remarked on the 'civility of the boatmen' on the Bridgewater Canal (but perhaps the duke was in earshot).

Bizarrely, there is no very clear evidence as to where the canal boatmen actually came from. One theory is that they were former navvies. Some possibly were – we don't know. Another idea, which was suggested in 1911, and took hold, was that the early 'boaters' were Gipsies. This idea doesn't surface at the time, though, and a study of early nineteenth-century censuses reveals very few of the more common Gipsy surnames amongst the boaters listed – nor their distinctive first names. Nevertheless the idea persisted until the 1940s.

It now seems more likely that the early boatmen were drawn from the ranks of rural labourers and possibly those of small farmers. Farmers very often operated in a 'dual economy', and they and their families were engaged in activities like spinning and weaving as well as farming per se. Another string to many farmers' bows was haulage – hiring out their horses and carts. This sideline as carters would have lent itself to the canal trade, which was of course basically a horse and waterborne cart.

It's unlikely the early boatmen were ex-mariners or river bargees. Some of them seem to have had a surprisingly gung-ho attitude to canal navigation. In the earliest days it was not uncommon for boatmen to leave the boat unmanned, so that it would bash from side to side, careering into lock gates and doing much damage. The canal companies were able to put a stop to this by passing bye-laws that insisted a 'steerer' always be aboard. It was pretty quickly realized that canal-boating was a two-person job.

Horse-drawn barge on the Leeds and Liverpool canal, 1900s

Often the second boatman was a boy. At first the man led the horse (suggesting that his background was on the land rather than the river) while the boy steered. Then it seems that the men asked themselves why they were doing all the hard work while the boy sat around and sunned himself on the boat. After the penny dropped the senior partner did the steering, and the junior walked the horse and worked the paddles, or sluices, on the lock.

Boatmen – sometimes encouraged by their employers – were notorious for trying to con the toll-keeper that they were carrying fewer goods than they actually were. Tolls were based on tonnage, mileage (hence the mile-stones on the canals, some of which still survive) and what cargo was being carried. Coal was very cheap to carry, as was culm (used for lime-burning) and limestone.

Finished goods were more expensive. The boater would have a ticket indicating how much tonnage he had on board. Of course, dishonest people had an interest in this paperwork being an underestimation: the carriers so that they could skimp on tolls, unscrupulous boatmen so that they could steal the excess. So the canal companies attempted to weigh the cargo as it passed through their toll stations. Boats were 'indexed' at special stations (one was at Etruria). Here weights would be lowered into an empty boat and metal strips called 'gauge indexes' fixed on to each side of the front and end of the boat to indicate how many tons were on board, depending where the waterline was in relation to the hull. The toll-collectors then worked out the weight of cargo by taking a reading from the index strips nearest to the water.

A refinement of this was to keep a record of every

boat's 'dry inches' and circulate it to the toll-collectors. A boat's 'dry inches', or 'freeboard', is the distance between the waterline and the gunwale, greater or lesser depending on how much cargo was on board.

The tonnage of any particular boat, then, could be worked out by a toll-collector, either by looking at the gauge index strips nearest the waterline, or by measuring the dry inches and looking them up in the vessel's 'gauge tables' in order to work out the tonnage being carried (the weight of the non-cargo contents of the boat having been subtracted). Often both methods were used.

Boatmen – or their bosses – soon realized that a false reading could be given by weighing down the boat on one side so that more dry inches would show up than were actually the case. From the 1830s four canal companies built weighing machines, and the whole boat was weighed, its known empty weight then being subtracted to arrive at an accurate total. One such weighing machine, originally on the Glamorganshire Canal, can still be seen at the Waterways Museum at Stoke Bruerne.

If the boat was carrying more than its way-bill pretended, a fine could be levied. What's more, the hefty sum of 2s 6d had to be paid to the gauger. The true figure was written on the ticket and so the boater – had he intended to do so – lost the opportunity of purloining any of the goods.

Where it was the company that was trying to pull a fast one and loading more cargo than was on the paperwork the boatmen – who were paid, per ton, by the amount on the ticket – were understandably aggrieved if they realized this was the case. One option was to simply throw the excess overboard. The Birmingham Canal, by 1795, was so full of jettisoned coal that boats were having difficulty getting through it.

Gauging cargo weights could take up a lot of time, of course, and often logjams of disgruntled boaters were formed outside the locks, waiting for it to be finished. Where the boatmen were violent, it was usually because what they thought were overly officious lock-keepers, toll-collectors or other canal employees were causing them to be delayed. They were paid by the trip, not by the hours they worked, or indeed waited. The canal companies regularly took action against or asked the carrying company to reprimand boatmen who had abused or assaulted their staff. Disputes with other boaters were, of course, not unknown.

The income of boatmen is hard to pin down and varied over time and place. But the senior man on board seems to have made a reasonable wage, compared to other working-class trades. That's not to say that they owned their own boats. Another misconception that grew up before the history of the canals was seriously studied, beginning in the 1940s, was that the majority of boats on the canals were owner operated, by skippers who became known as 'Number Ones'.

Boats represented a serious capital outlay beyond the means of most of the working class. Although by dint of extremely hard work, canny footwork, luck or perhaps a family legacy some boaters did own their own boats, by far the greater number were owned by wealthy capitalists. Collieries, iron works and manufacturers very often purchased their own fleets. On a greater scale were the carrying companies, such as that of Hugh Henshall and Co. on the

Trent and Mersey Canal. Inventories have shown that the great majority of boats were owned by individuals or businesses who had an average of six boats each. Those owners with fewer than six boats were most often businessmen, sometimes banding together, who had bought them as a means to aid their own trade, or as an investment.

The singleton owner operator was a rare bird. Perhaps 4 per cent of the canal boats were owned by 'Number Ones', and, although the proportion varied from canal to canal, it seems unlikely that this figure ever reached more than 6 per cent throughout the whole commercial canal era. For the most part, throughout their history, trade on the canals was dominated by the larger concerns, who could pick and choose the higher-value cargoes. The smaller operators tended to carry those that attracted the lowest fees, coal being pre-eminent, along with limestone, straw, hay and other high-volume materials. The small operator at the bottom of the pile had also to carry the most unpleasant cargoes to handle, such as manure (including human waste, or 'night soil') and bones. Such loads brought with them infestations of maggots and vermin. When we later hear of boats that were in a very dirty condition, it was usually these sorts of vessels, working at the least profitable and filthiest end of the market, which were being referred to. (Regular fumigations, or 'bugging outs', were necessary on every canal boat.)

The owner-operator was also more vulnerable because of lack of the capital needed to keep his boat in regular repair. If calamity struck, such as a falling off of trade, the death of his horse, illness or a long period frozen up, the small man didn't have the resources to weather the storm and often went to the wall.

The American War of Independence, from 1775 to 1783, was a financial disaster for Britain. Prices rose, while the availability of capital fell. The result of this was that canals that hadn't been finished by the time of the American Revolution were left high and dry.

Two of the canals making up Brindley's Grand Cross were stymied. The Coventry Canal, which began construction in 1768, didn't reach the Trent and Mersey, as planned, until 1789. The Oxford Canal had started running out of cash even before the war started, and the Thames wasn't joined until 1790, twenty-one years after the canal had been begun. This was despite the fact that the other canals, rather eager to have a link to London built, strongly pressed this dysfunctional and complacent company to get a move on.

(There had already been a richly absurd development. When the Oxford Canal met the Coventry Canal, near Hawkesbury, in 1771, the two companies couldn't reach an amicable agreement on tolls, and so for a mile they ran in parallel, within hailing distance of each other, until the Coventry Canal won an injunction, five years later, and the shadow Oxford branch was eliminated.)

But at last Brindley's vision was realized.

The Grand Union Canal

Introduction by John Sergeant

The canal system's waterways were vital arteries spreading across the country. It also needed a backbone linking London to Birmingham, and that was provided by the Grand Union Canal. As you potter gently along this delightful reminder of an earlier age there can be a sudden whoosh from a speeding train on the line nearby. Each transport age has had to find the most efficient way to move goods and people between the capital and England's second city. Not surprisingly the canal, the railway and the motorway often took the same course.

The main route from Birmingham used to join up with the Oxford Canal and the River Thames to take boats to London. The biggest building project of the time, and one of the most ambitious canal extensions ever undertaken, shortened the journey by 60 miles. The new canal, completed in 1805, starts at Braunston, close to Daventry; and under the arch of a beautiful iron bridge our journey to London began. I quickly saw how the importance of the waterway allowed the engineers to spend vast sums overcoming huge problems.

Two mighty tunnels had to be driven through neighbouring hills, and both are without towpaths. Going through the first, one of our experts kindly demonstrated the rough art of legging. He stretched out on a board and used his feet to drive the narrow boat through the darkness. I played a less heroic role, sitting on the other end of the wooden plank. At the entrance to the second tunnel, at Blisworth, I walked up the track the horses would take to meet their boats on the other side. Only the large brickwork chimneys of the ventilation shafts give any hint of the two mile-long tunnel many feet below.

I appreciated more fully the skill of the canal-builders. Tunnels offer a more direct route and they avoid time-consuming locks taking the boats up and down hills. They can also be a way of keeping the canal below the level at which water normally settles. Canals above the water table will run dry without a reservoir or some form of pumping system. There is more to canal building than meets the eye.

On the Grand Union it is easy to forget history and simply enjoy a summer cruise through some of the most attractive canal scenery you can find. The canal is wide along most of the route, because traffic was heavy. Willow trees dip into the water; there is abundant wildlife; and fish jumping out of the water are an obvious sign of good conservation. When you sometimes hear the noise from trains rushing along the main line it simply adds to the feeling that you have made the right transport choice.

Closer to London the mood changes quite quickly. When you cross the aqueduct over the North Circular, it is still gratifying to see cars jamming the road below. But ahead the canal takes on a more subdued air. The fields have given way to industrial sites. Some of the bridges are daubed with graffiti, and the general lack of canal traffic encourages green algae to spread across the water.

Here and there bold attempts are being made to cheer up the scene, and the mood changes again as you approach the centre of the city. We glide through posh Little Venice, and coming into Paddington Basin there is an explosion of new buildings. Sparkling high-rise offices and fashionable restaurants form a brilliant end to the Grand Union Canal. If you have the time, and the inclination, this is by far the best way to travel from Birmingham to London.

The Grand Union Canal has been, since 1927, an amalgam of existing waterways and the Grand Junction Canal. It was put together by the Regent's Canal Company in a bid, somewhat late in the day, to create a functioning network of canals rather than a viper's nest of competing companies, each looking after its own interests.

The nomenclature is somewhat confusing. In 1894 the Grand Junction Company bought what was then called the Grand Union Canal, which makes its way from Norton Junction, near Daventry, up to Foxton and Market Harborough. This stretch of waterway is now known as the Old Grand Union – or, to ring the changes, the Grand Union (Old). The Leicestershire and Northampton Union Canal, which links to the Grand Union (Old), and on towards Leicester itself, was also purchased.

In 1927 this whole waterway became part of the Grand Union Canal (incorporated in 1929), when the Regent's Canal bought the Grand Junction and its offshoots, as well as the three canals linking it to the heart of Birmingham – the Warwick and Napton, Warwick and Birmingham, and Birmingham and Warwick Junction Canals.

In 1932 the Grand Union Company also bought the canals that joined up with the northern end of the Leicestershire and Northamptonshire Union – the Leicester and

Loughborough Navigations and the Erewash Canal. This gave the Grand Union a through route from Limehouse Basin, on the Thames, up to central Birmingham on one spur, and to Leicester and the East Midlands coalfields on the other.

Regent's Canal at Paddington Basin, London

The Grand Union Canal had also tried to buy the Oxford and Coventry Canals, but they refused to sell and remained independent.

The Grand Junction Canal, authorized in 1793, completed in 1805, was, once built, a highly successful venture and remained a mainstay of the whole commercial network until the 1960s.

The logic behind the new canal was unarguable. The Oxford Canal had finally been finished in 1790, but the route offered to London from Birmingham by the last part of Brindley's Grand Cross was doubly circuitous. In the first place the journey involved the absurdity of travelling north-east from central Birmingham, along the Birmingham and Fazeley Canal up to the Coventry Canal. Coming south from there to London, via a wide arc through Oxford, before joining the Thames, also made little sense. What's more, the journey was intrinsically slow, thanks to the leisurely curves of Brindley's contours. The route from central Birmingham to London Bridge through the Grand Cross was more than 230 miles, and yet only 105 miles as the crow flies.

In 1791 an alternative was proposed, a canal that would run south-east from Braunston, on the Oxford Canal, down to join the Thames at Isleworth. (Brentford was later chosen as the terminus.)

A survey was carried out at the expense of the Marquis of Buckingham, who seems, in the spirit of Wedgwood and Bridgewater, to have been more interested in the canal as a stimulus to trade and employment in the hinterland it passed through than as a money-making venture in itself. The survey was carried out by James Barnes, who had worked on the Oxford Canal. This was, eventually, to become the Grand Junction Canal.

In 1792 a public meeting to promote the new canal was held at the Bull Inn, Stony Stratford, and the plans were given a rousing thumbs-up. William Praed, MP, was elected as chairman of the committee, which now tried to get an Act of Parliament passed in its favour. (He is remembered in Praed Street, Paddington, near to where that branch of the Grand Junction terminates.)

The promoters now drummed up support for their plans. A more detailed survey was carried out by Barnes, now under the supervision of William Jessop, who was to become the supervising engineer. The committee cannily offered landowners whose properties would be carved through en route the chance to buy shares, the number being dependent on how many miles of canal would pass through their estates. At this stage it was decided that the canal would take a southern route to Uxbridge, where it would join the River Brent, which would be made navigable down to the Thames at Brentford.

The Oxford Canal, horrified at the idea of losing its traffic south of Braunston, strenuously opposed the new canal. Hardly able to argue that the existing route through Oxford was ideal, in 1792 it suggested a new canal of its own devising. This would, naturally, use as much of the Oxford Canal as it thought it could get away with. Initially styled the 'London and Western Canal', it was projected to run from Hampton Gay, near Thrupp, to Marylebone. This proposed waterway – whose route changed on several occasions – became known as the 'Hampton Gay Canal'.

It was clear that if both canals went ahead they would destroy each other, there being insufficient traffic to support both. The Marquis of Buckingham tried to broker a compromise with the two groups, suggesting that the Oxford Canal build a line to Marsworth, where it could join the Grand Junction. The latter company was having none of this, and a bad-tempered stalemate was reached. Both parties prepared a Bill, and the proprietors of the Oxford and Hampton Gay Canals furiously lobbied against the *arriviste*.

But the Grand Junction management was a ritzy set-up. Along with the support of the Marquis of Buckingham they could count on that of several powerful landowners whose estates the canal would cut through, including the Duke of Grafton and no less than five earls. With the help of their political clout the Hampton Gay Canal was seen off, leaving only the Oxford Canal to be dealt with.

A compromise was finally reached. The new canal would guarantee the income of the old. Discussions were also held to lease the section of the Oxford Canal from Braunston to Hawkesbury, where it joined the Coventry Canal.

But a far better scheme had by now been proposed. This was a plan to extend the Warwick and Birmingham Canal down to Braunston to meet the Grand Junction. Not only would this have provided the most direct route into and out of the town. It would also cut the perennially troublesome Oxford Canal Company out of the main route from Birmingham to London altogether.

Sadly, the Oxford Canal, after much ear-bending behind the scenes, managed to get this new canal diverted from its intended target of Braunston to Napton, on its own canal. (This new section was then called the Warwick and Napton Canal.) The result of this was that the Grand Junction, in order to get to Birmingham, had to lease five miles of the Oxford Canal. The main line from London to Birmingham would thus be via the Grand Junction, Oxford, Warwick and Napton, and Warwick and Birmingham Canals, with the eastern part of the city still served by the Oxford, Coventry, and Birmingham and Fazeley Canals.

Job done, the Oxford promptly put its tolls up on this section. The canal would be a thorn in the side of the Grand Junction, later Grand Union, for decades to come.

So too would be the question of gauge. The Grand Junction was to be a broad canal, capable of taking 70-ton barges, and its owners manfully tried to persuade the Oxford, Coventry, Birmingham, and Trent and Mersey concerns to widen their canals too. Broad canals were the future. But the narrow canal companies were rooted in the past, and despite many attempts on the part of the Grand Junction to persuade them to undertake the (admittedly enormous) expense of widening their canals, the older companies never did.

In 1793 the Grand Junction Canal received its Authorizing Act, and work went ahead. William Jessop was the chief engineer, and James Barnes the resident, or full-time, engineer in charge of construction.

Unlike James Brindley's canals, the Grand Junction was going to have no truck with meandering contours but would blaze a trail as directly as it could. Its length would be 135 miles – from Brentford to Braunston would be 93 miles (with a further 35 miles to Birmingham). This would mean some difficult engineering. At the main summit, at Tring (Braunston was the other), a huge cutting of 30 feet was to be made. Long tunnels would have to be built at Braunston and Blisworth. (A third tunnel was projected, but never built.) The cutting and tunnels meant fewer locks were needed, and reduced the neccessity to pump water up to a higher level, which would have been expensive. But they were considerable undertakings.

A horse hauling a barge on the Regent's Canal

Work was far from straightforward. As with most canals, once the actual construction began problems were discovered with the underlying terrain, which caused delays and added expense, especially if a deviation from the plan deposited with parliament had to be taken. The Authorizing Act gave the company the right of compulsory purchase of any land within 100 feet of its proposed route (disputes to be settled by independent commissioners). If it had to change this route, all bets were off, and the company had to negotiate with the landowners afresh.

This was the case with the Earls of Essex and Clarendon, who, even though they both sat on the company's board, extracted a high price for the canal to stray into their estates.

Problems with securing labour, and bricks, were also to cause delays. But one of the main difficulties was self-inflicted. Because of the dilatory approach of some of the contractors who had been hired to build the individual sections and/or those subcontracted by them, much shoddy workmanship was carried out. In 1806 this led to a section of the embankment for the aqueduct that crossed the Great Ouse at Cosgrove to collapse. Worse was to come. Two years later the whole thing gave way and was only just prevented from flooding the neighbouring area. Much money was wasted putting right this and other badly fashioned constructions.

Further problems emerged at Braunston Tunnel, where the two ends of the tunnel went off course, meaning that the tunnel has a distinct kink.

But by 1796 it was finished, and by 1800 most of the line was complete.

The Blisworth Tunnel, however, was causing even more headaches, taking some twelve years to build – as opposed to the original estimate of four. It had to be driven through clay, rock, limestone and ironstone, and the ground had a disturbing tendency to move, especially in wet weather.

Flooding in the tunnel caused incessant difficulties. At one point Jessop wanted to throw in the towel and build a series of locks instead, but Barnes persuaded him to keep faith. John Rennie and Robert Whitworth were called in as consultants, and Rennie solved the issue by suggesting a new line to intersect the one already built, and the construction of a small tunnel below, to draw off the water.

A double tramroad, made of cast-iron rails, was constructed by Jessop's partner Benjamin Outram, the future tramroad king. This took traffic between the two canal sections while work continued on the tunnel. (The cargo had to be laboriously unloaded and reloaded.)

To help conserve water, in 1803, near Berkhamsted, the Grand Junction pioneered the use of side ponds. These were built adjacent to the lock. When the lock was emptied, half the water was siphoned off, whereupon it would feed back into the next lock, saving water. Most are now out of service – never having been all that efficient.

Water supply continued to pose a problem. At the four-flight locks at Kings Langley, the canals' old enemy, the millers, in this case owners of paper mills, were complaining that their water was being reduced. This was tackled, at great expense, by the use of Boulton and Watt steam engines for 'back-pumping'. But this procedure didn't solve the problem to the millers' liking, and so the company built several side ponds. They didn't work either, and with nothing else able to satisfy the litigious millers, in 1818 this troublesome section of the canal had to be diverted, and new locks built. Thomas Telford supervised the construction. Other millers, meanwhile, put a spoke in the wheel elsewhere and had to be mollified or bought off.

William Jessop (1745–1814), engineer

The extra costs incurred because of these engineering problems and delays, as well as rising prices, meant that the canal committee was almost constantly having to raise extra money, through loans and new subscriptions. But it managed to keep going. (The nominal capital controlled by the firm made it one of the largest in the whole world.)

Branches were built to Buckingham, Newport Pagnell, Aylesbury and Wendover (one of the sources of water for the main line). At Northampton the canal joins the Nene Navigation, which connects via the Fens to the Wash.

The Paddington Branch had first been surveyed, by Barnes and Jessop, in 1794. It was opened in 1801. It may have been something of an afterthought, but it was an extremely good one. The Paddington Branch stemmed from Bulls Bridge, at Southall, running north-east then due east through flat countryside (as much of it then was) and brickfields, and on to Paddington, requiring no locks to be built.

Although much traffic was carried on from Bulls Bridge south-east to the Thames at Brentford, the Paddington Branch was a vital one, especially after it linked up with the Regent's Canal, finished in 1820 (the same canal that was to set the Grand Union scheme in motion a century or more later).

From Little Venice, near Paddington Basin, the Regent's Canal follows an eastwards course through Regent's Park, Camden, Islington and Hackney, before heading south through Bethnal Green to join the Thames at the Regent's Dock, more commonly known as Limehouse Basin. Though the dock couldn't handle the ever-larger sea-going ships, it could accommodate coastal vessels and became an important source for the Regent's and Grand Junction's trade, especially in coal, and later steel. Thus a highly important link from the Pool of London to Birmingham, and beyond to the Potteries and Mersey, had been made.

Entrance to the Regent's Canal at Limehouse, 1827

There was one problem. Under the terms of the Grand Junction's Authorizing Act, which the importers of sea-coal from Newcastle to London had managed to influence, although coal could be carried on to the canal from Regent's Dock, it couldn't be brought in from anywhere north of Hertfordshire without the payment of special tolls to the sea-coal cartel, if more than a certain level of tonnage came in. But these restrictions were gradually lifted, and coal became the main trade of the canal.

Despite these annoyances, the Grand Junction Canal

was an immediate success. So much so that water supply became a problem, and more reservoirs had to be built, and water pumped to supply the Tring summit. Over the course of the nineteenth century, additional reservoirs were built on a regular basis, gradually increasing the supply of water.

Though barges of up to 14 feet wide operated on the Paddington and Brentford Branches, and below Berkhamsted, north from there most of the craft using the canal were narrow boats.

To speed up traffic, duplicate locks, which seem to have started on the Regent's Canal, were built on the Grand Junction from 1835, when seven broad duplicate locks were constructed beside the flight at Stoke Bruerne. (In some duplicate locks, paddles allow the water to drain into the adjoining chamber, similar to the way side ponds operate.) From Marsworth to Stoke Hammond a series of narrow-gauge duplicates were built, allowing single narrow boats to pass through. (Before this they had been discouraged from using one whole broad lock just for themselves, unless water supplies were buoyant, but had to wait for a companion to show up.)

When more reservoirs were built, and the water supply improved, duplicate locks fell out of favour, and most were filled in. (What seems to have been the prototype duplicate lock is still to be seen at Hampstead Road Lock, at Camden Market, on the Regent's Canal.)

Despite the Grand Junction's stated distaste for narrow boats, their presence on the canal actually worked fairly well, because two narrow boats could fit into one broad lock. This became of immense value when narrow boats began working in pairs – they could 'breast up' in the locks rather than having to uncouple.

Duplicate locks were introduced because, in 1838, what was to be the Grand Junction's main rival, the London and Birmingham Railway, opened for business. Its threat was hard to ignore, as it ran right past the canal for several long sections and crossed it six times. The Grand Junction having rarely carried passengers, that trade wasn't affected, but the fly-boat trade was badly hit. Heavier cargoes, though their tolls had to be drastically reduced, carried on in the same tonnages as before. (More and more coal was being produced. But much of the surplus now went to the railways.)

A duplicate lock on the Regent's Canal, *c.* 1850

In 1848, as a response to railway competition, the Grand Junction began operating its own carrying trade. In 1864 a fleet of steam narrow boats was put into operation, and steam tugs were introduced at Braunston and Blisworth in 1871.

The carrying company was wound up in 1876 after a famous incident at Macclesfield Bridge, on the Regent's Canal, two years before. One of the Grand Junction Carrying Company boats, transporting, perhaps a little unwisely, a cargo of petroleum and gunpowder, blew up, taking out the crew, boat and bridge and damaging buildings and windows at a mile's distance. (Railway companies refused to handle such combustible materials.) When the compensation claims were added up, the total came to an eye-watering £80,000, which the carrying company were on the hook for. The bridge became known to boaters ever after as 'Blow Up Bridge'.

The important firm of carriers Fellows, Morton and Clayton took over much of the trade. Partly at its behest, in 1894 the Grand Junction bought the Grand Union Canal (Old), which led up to the important coalfields of Leicestershire, and thence to join the Trent. Fellows, Morton and Clayton also encouraged them to build wide locks on this stretch of waterway, promising to do its bit by employing barges on the canal. The idea wasn't followed up.

But, in 1900, the Grand Junction did experiment with 'inclined planes' alongside the flight of ten locks at Foxton, at the top of the old Grand Union line. (Significantly, these locks, in two staircases of five, were narrow gauge, the company having given up on the idea of persuading the other narrow canals to reinvent themselves as broad ones.)

A boat being carried uphill on an inclined plane, 1880

Two caissons, or watertight containers, counterbalancing each other, ran on carriages, in opposite directions, up and down rails built into the slope. They were capable of carrying one barge or two narrow boats side by side. The basic idea had been pioneered at the Ketley Canal in Shropshire, in 1788, following an earlier design of John Smeaton's. Here, boats were passed along a slope, over a series of rollers, a laden boat pulling the empty boat up.

Inclined planes were much discussed from the 1870s, when improvements were sought to try to fight back against the domination of the railways. The time-consuming locks – especially narrow locks – put the canals at a considerable disadvantage, and planes seemed to offer an answer. But they were never a success, and at Foxton the incline had been abandoned by 1910. A similar arrangement meant for locks near Watford was never put into place, when it became clear that the planes were un-economic, there not being enough traffic to support the cost of operating them.

But traffic on the canal, both in terms of the number of craft and the amount of tonnage, continued to increase, peaking in 1914. After the war, unlike most canals, the Grand Junction was still doing well, although traffic was declining. In an attempt to shore up the network, in 1927 the Grand Union Canal (New) was put into place. In 1934 it bought its own carrying fleet, which was to become the largest in Britain, including 185 pairs of narrow boats.

Starting in 1932, backed by government money made available to relieve unemployment, the company embarked on an ambitious series of works to widen fifty-one locks between Braunston and Birmingham, along with several

The Grand Union Canal –
The journey itself

Moorings at Watford

In his *Tour of the Grand Junction Navigation*, published in 1819, before the Railway Age, the writer and water-colourist John Hassell had this to say of the waterway:

The Grand Junction or Braunston Canal is so peculiarly distinguished, that truly it may be said, from its junction with the Thames to its termination at Braunston, to be an almost perpetual succession of variegated beauty, shaping its devious course through some of the richest vallies of Middlesex, Hertfordshire, Buckinghamshire and Northamptonshire, accompanied by an abundance of the most luxuriant scenery, and lined on its sides with a succession of rising eminences.

There are still many beautiful stretches, as the canal makes its way past what are now urban and industrial areas, cleverly confining itself to the countryside where it can. But the canal was not, of course, conceived for leisure. It was first and foremost a commercial proposition, the major section of the vital trade route from London to Birmingham. The journey we'll be making is the 87 miles from Braunston, the northern end of the Grand Junction, to Bulls Bridge, Southall. We then travel the 13 miles of the Paddington Branch to Little Venice.

Braunston, in Northamptonshire, was, and remains, one of the most important hubs of the British canal network. Michael Pearson, in his 'Canal Companion' series, remarks that 'Braunston symbolises the magnetism of the Midland canals, and is a point of pilgrimage which has captured the imagination of waterway writers, artists and photographers more than almost any other canal location.'

Braunston Turn – the junction with the Oxford Canal – is Y-shaped, with a small triangular island under two fine bridges made by Horseley Iron Works. Opposite is the first of many canalside pubs, the **Boat House Hotel**, a former mill. **Braunston village** is to the north and east. It is dominated by the nineteenth-century **All Saints**. (A church has been on the site for a thousand years.) Of note too is a manor house and a sail-less red-brick, castellated windmill.

On the canal itself, as we head to what was the boatyard and dock area, now a marina, is the **Stop House**. It was built in 1796 so the Oxford Canal Company could collect its exorbitant tolls. It may seem like an odd place to put it, until one realizes that the original line of the Oxford Canal at Braunston included yet another of Brindley's curves, and it originally joined the Grand Junction where the marina is now situated. The new, straighter arrangement dates from the 1830s. The marina is divided by **Brindley Quay**, at the end of which is the handsome brick **Butcher's Bridge**. Another Horseley bridge is to be found at its western end.

The marina was once the home of **Nursers** boatyard, famous for its paintings of roses and castles on narrow boats. One of the last carriers, Blue Line, was also based here. In the canal's commercial heyday Pickfords, and later Fellows, Morton and Clayton, had important depots at Braunston, as did the coal carriers Samuel Barton and Co. Willow Wren, formed after the Second World War, managed to stay in the long-distance carrying trade until 1969. At the **Bottom Lock** the old forge is now a boatyard. Nearby is a pump house dating from 1897. The famous canalside pub the **Admiral Nelson** is in a building that pre-dates the canal.

Leaving the canal village, we enter **Braunston Tunnel**, one of the two summit levels. It was built out of alignment, with a kink in it. It's 2,042 yards long, and took three years to finish. It's now the seventh-longest navigable canal tunnel in the country. Two narrow boats can pass.

Emerging from the tunnel, we're soon at **Norton Junction**, where there is a connection with the Grand Union (Old). This canal passes through Watford Gap on its way to join the Leicestershire and Northamptonshire Canal at Foxton, near Market Harborough. The **toll-house** here used to be the base of a Salvation Army commander, who, with his wife, ran two mission boats along the canal in the 1950s, ministering to the boat families.

We begin to descend into a valley, with views of the Northamptonshire Uplands, arriving at the canalside **New Inn**, adjacent to the lock at **Long Buckby Wharf**. Buckby was once famous for its shop selling the traditional brightly coloured water cans, still a prized feature of restored narrow boats. The old Roman road of Watling Street, mostly now the A5, is from now on to be an occasional companion en route.

At **Weedon Bec** a Royal Pavilion was built in the early 1800s, intended not for the Prince Regent, but for his father, George III. Chosen because it was pretty much in the centre of England – and linked to London by canal – it was to become the last stronghold of the monarch in the event of an invasion by Napoleon. (Who was indeed planning to do just that, mustering his troops on the cliffs at Boulogne in 1803.) Eight large storehouses were built, plus a barracks, and the complex, which manufactured small arms, became known as the **Military Ordnance Depot**. It was served by its own feeder canal, which passed under a portcullis to show that the army wasn't messing about. It's been disused since the 1920s (though it's thought that had Hitler succeeded where Napoleon had failed, the Princesses Elizabeth and Margaret would have been taken here, and then flown to Canada).

From **Flore Lane Bridge** the canal winds through a pleasant, rural area. We come to **Bugbrooke**, a pretty village near the canal, with a fine thirteenth-century church, **St Michael and All Angels**, and the **Wharf**, a canalside pub. We then pass through agricultural land, past mileposts marked 'GJCCo', to **Gayton Junction**, where the **Northampton Arm** begins. A visitor to this branch, after grappling with seventeen locks, will find a renownedly beautiful waterway.

On the main line, we come to the attractive village of **Blis-** **worth**, once a busy canal centre, where the Grand Union Canal Company operated several warehouses, one of which still stands. Just before the village is **Candle Bridge**, so-called because from a cottage nearby a woman sold tallow candles to the leggers who worked the notorious **Blisworth Tunnel**, which burrows through a ridge of ironstone to connect Blisworth with Stoke Bruerne. The tunnel, twelve years in the making, was one of the hardest to construct of all British canal tunnels. It had to penetrate layers of rotten oolite and heavy clays, which were constantly being flooded by natural springs. Today it's the third-longest canal tunnel in operation (the seventh-longest at the time John Hassell was writing).

While the tunnel was being built, cargoes had to be taken by horse across the top by means of a double iron tramway. Remains of the tracks are still visible, although the ironwork was dug up and used on uncompleted sections of the Northampton Branch. Finally finished in 1805, the tunnel, 3,056 yards long, had no towpath, and thus boats had to be legged through. Having to scuffle boats along for more than a mile and a half, the leggers certainly earned their crust – until steam tugs took over, in 1871.

Before this, steam-powered vessels had already been in operation. Their early use of the tunnel was not without incident. In 1861, onlookers were astonished to see a steamer slowly emerge at Stoke Bruerne with, apparently, no one steering, or indeed aboard. It turned out that in the narrow confines of the tunnel dense smoke had been blown back into the boat from the engine, and the crew had been asphyxiated. Two men were revived, while two others were found to be dead. Ventilation in the tunnel was afterwards increased. Distinctive brick ventilation shafts are still to be seen on the path over the top. They became infamous for cascading water on to unsuspecting boaters.

At the other end of the tunnel is one of the most famous of all inland waterway locations, **Stoke Bruerne**. An 'ideal canal village' as the Collins and Nicholson Guide puts it. As at Braunston, it was a popular choice of 'land home' for retired boaters and their families. Many original buildings survive. At the **Boat Inn** (still thatched) boat crews waited for the leggers to bring their boats through – while exhausted leggers, presumably, slaked their thirst. The **Canal Museum**, the first of what are now several museums devoted to inland waterways, has been here since 1963, situated in a former corn mill.

Duplicate locks were added to the seven at Stoke Bruerne in 1835. They didn't really aid matters and were soon filled in. The one that survives is now the home of an original boat-weighing machine, one of only three that were built, from the Glamorganshire Canal.

One of the most famous of Stoke Bruerne's residents was Sister Mary Ward, who, while not formally trained as a nurse, from the 1930s to the 1950s voluntarily dedicated herself to the care of the boat families and became their principal source of medical aid, often buying the medicine herself. Her 'surgery' was at the **Stoke Top Lock**. Towards the second of the five Stoke Locks is the **Navigation Inn**, built in 1822. At the bottom of the flight is a pumping house.

All the side ponds have been filled in.

We now pass over the River Tove, via an aqueduct, and travel through remote countryside and pastureland along the valley, under several brick 'accommodation' bridges, to the strangely named **Yardley Gobion** (Hugh Gobion was a twelfth-century landowner) and then on towards the Ouse Valley.

Soloman's Ornamental Bridge, a Gothic construction of 1800, precedes Cosgrove. At **Cosgrove Lock** the branch to Stony Stratford and Buckingham once spun off. It's now disused. Leaving the village, the canal passes over the Great Ouse by means of the ill-fated **Cosgrove Aqueduct**. After its collapse in 1808, William Jessop designed an iron trough to connect the two embankments, completed three years later. At Cosgrove the iron towpath is cantilevered over the water, unlike Pontcysylte, where it rests on iron pillars. A 'cattle creep' cuts through each arm of the supporting masonry, allowing farm animals, horses or humans to pass between. Traces of one of the temporary locks built while the aqueduct was first being constructed still remain. It marks the border between Northamptonshire and Buckinghamshire.

The village of **Wolverton** became an important railway centre in the nineteenth century, and the Royal Train was built here. There was once a canal branch to Newport Pagnell, later turned into a railway line. After the urban hiatus at Wolverton, we head towards the Ouzel Valley, along a wooded section of the canal, with the attractive **Linford Park** to the south (its mansion house a noted recording studio). Then we are in the new city of **Milton Keynes**. Near the Open University campus at Woughton Park, the ghosts of Dick Turpin and Black Bess are said to stalk **Woughton on the Green**. The pretty village of **Simpson** is at the edge of Milton Keynes, with its canalside **Plough** inn.

From here the canal begins its ascent over the Chiltern Hills. And yet its first step is somewhat faltering. **Fenny Lock**, at Fenny Stratford, has a rise of only 13 inches, leading to a long-standing legend that this was due to a mistake on the part of the surveyors, one of whom, overcome with mortification, then drowned himself in the canal. The more prosaic explanation is that this was quite deliberate, the engineer's hand being forced by difficulties in equalizing the levels of the canal on either side due to flooding. The steep hills of Bletchley, home to the 'Ultra' decoders in the Second World War, are to the west as we cross Watling Street and continue through the Ouzel Valley.

After the attractive village of **Stoke Hammond** comes the 'Stoke Hammond Three' – not a group of prisoners of conscience, but a flight of locks at **Soulbury**. One of the original 'northern engines' (by which is meant, from the northern end of the Grand Junction) pump buildings is by the side of the lock, as is the **Three Locks** pub. Woburn Abbey is some five miles away. At **Old Linslade** Buckinghamshire gives way to Bedfordshire.

Although the duplicate narrow locks between here and Marsworth were filled in, the duplicate narrow arches built into the bridges remain.

We pass through peaceful, remote grassland to **Leighton Buzzard** – perhaps a corruption of the more Norman-sounding 'Beaudesert'. Here can be seen the traces of old duplicate locks, as well as engine houses, wharves and original lock-keeper's cottages. The boatyard of L. B. Faulkner here was also well known for its painting of canal boat decorations. A disused narrow-gauge railway runs right by the canal south of **Linslade**. From here all the way down to Tring is one of the most celebrated stretches of the Grand Union Canal.

Towards **Ivinghoe**, with its prominent beacon, we come to the **Chiltern Hills**, with their fine beech woods. A chalk lion is to be seen from the canal, an advertisement for nearby Whipsnade Zoo, dating from 1933. To the south is Cheddington. At the Bridego Bridge nearby, the Great Train Robbery took place in 1963. At **Seabrook Locks** is a well-preserved back-pumping 'northern engine'.

The canal now girds its loins for its ascent towards the summit at Tring. At **Marsworth Junction** is the **Aylesbury Branch**, a narrow canal. The seven attractive **Marsworth Locks** were known to generations of boaters as 'Maffers'. The duplicate at **Marsworth Top Lock** was converted into a dry dock, which is still there. The bridge nearby is another with a second, smaller arch. The locks are surrounded by superb countryside and a nature reserve.

To the south are the three reservoirs needed to feed the summit, built between 1806 and 1817. Water was pumped up by steam. At **Bulbourne Junction** is the **Wendover Branch**, constructed mainly to supply water for the main line. At **Bulbourne**, workshops, an old forge and wharves still survive. Lock gates were built at the **Bulbourne Works** until recent times. The building remains, complete with an Italianate water tower and ornate weather vane.

Bulbourne marks the transition into Hertfordshire and the start of the 1½-mile **Tring Cutting**. It was dug 30 feet deep. This was a huge undertaking for its day. Earth was carried away in barrows, up wooden runways, assisted by pulleys using horsepower. As they didn't know what else to do with the earth, some of it was added to the banks, making the cut look deeper than it actually is. The summit is nearly 400 feet above sea level. The **Tring Summit** itself is three miles long.

Near **Aldbury** there is an elegant column commemorating the 3rd Duke of Bridgewater, the great canal pioneer, who owned the beautiful Ashridge Estate, to the east. The summit ends at the wonderfully named **Cowroast Lock**. John Hassell concludes that this is a corruption of 'Cow Rest', dating from the time when cattle were taken by 'drove' road to London. Another ancient path, the **Ridgeway**, is nearby. Near **Northchurch** top lock, **St Mary's** is one of the world's oldest flint-built churchs.

In days gone by, watercress, from this section of the canal, was taken down to the busy inland port of **Berkhamsted** – known in boating patois as 'Berko'. There is little trace of the eight busy wharves that were once in operation here. But there are churches and coaching inns, as well as Berkhamsted Castle, which dates from the eleventh century, and Berkhamsted School, from the sixteenth. Some of the many traditional canalside pubs are still in business, such as the **Boat** and the **Rising Sun** – found near the lock of the same name. This and **Berkhamsted Bottom Lock** were known as 'Sweeps', after Eli Oliffe, a chimney sweep who kept a boatman's store nearby. Berkhamsted was the northern limit for barges – as opposed to narrow boats – on the Grand Junction Canal.

Passing out of Berkhamsted, we come to **Bourne End**. **Irishman's Lock** here commemorates Joseph Buck, a lock-keeper who drowned on Christmas Day, 1898. The canal now reaches **Boxmoor**, on the outskirts of Hemel Hempstead, and the attractive **Fishery Inn**. Up until 1981, Rose's lime cordial factory here received its shipments of lime juice by canal (it was known as the 'barrel run').

Leaving the Chilterns behind, we come to a section of canal dominated, until recently, by paper mills, which were supplied with coal from the canal and shipped their goods along it. Two of them, at **Apsley** (where an old warehouse stands) and **Nash** (which closed in 2006), were owned by the very sizeable concern of John Dickinson Sta-

tionery Ltd (famous for its 'Lion' brand), which operated its own fleet of boats – called paper mill 'dashers'. Ironically, the man who founded the company was one of the millers who caused the Grand Junction Company such trouble back in the early 1800s, meaning that the canal's course here had to be diverted. **Frogmore Mill** still survives. Built in 1803, it was the first mechanized paper mill in the world.

The **Ovaltine factory**, which also owned its own fleet of boats, operated at **Kings Langley** until it closed in 2002. The façade facing away from the canal is a typical example of late 1920s factory Art Deco. We now come to the Gade Valley.

Lady Capel's Lock was once the point at which coal coming south had to pay a separate toll. Naturally, coal was unloaded north of here to avoid having to cough up. Lower down, the canal's original route was changed to avoid the expense of a tunnel at Langleybury, and permission was sought from the two landowners, the Earls of Clarendon and Essex, to divert it through their lands. John Hassell takes a very kindly view of their motivation in allowing this to happen, saying: 'it must stand as a monumental record, and example, of the urbanity and *amor patriae*, these distinguished noblemen exhibited for the weal of their country'. Love of money may also have come into it. Both were on the canal company's board and both extracted a hefty price. They also insisted that the canal had to mind its manners as it passed through their properties. This makes this section of the canal a particularly attractive one. At **The Grove**, owned by the Earl of Clarendon, there is a fine stone bridge, a Grade II listed building.

As it reaches **Cassiobury Park**, on the outskirts of Watford, the Grand Junction becomes even more playful. Not only is there another elegant bridge, with a neo-classical balustrade. The Earl of Essex also insisted that the canal and its four locks become part of the park, gently undulating through it so as to form a feature of the landscaped wooded parkland. This is a fine part of the waterway. According to Hugh McKnight, in the *Shell Book of Inland Waterways*, it's 'one of the loveliest sections of canal so near a town anywhere in Britain'.

At **Croxley** there used to stand the vast Croxley Paper Mills, the third of John Dickinson's three great mills, which was supplied with coal by the canal until 1970.

The canal now approaches the water town of **Rickmansworth**, where the rivers Gade, Chess and Colne all meet the canal, with disused gravel pits (one of them being the American-sounding 'Rickmansworth Aquadrome') forming attractive lakes. The town, known to boaters as 'Ricky', was in the canal's heyday a thriving centre of water-bound trade, and an important centre of boat-building. The craft produced, inevitably, being called 'Rickys'. Along with a number of wharves, depots and yards there was a railway interchange basin. **Frogmoor Wharf** was once the site of the well-known boatbuilders W. H. Walker and Sons. Their last two boats were made in 1952. Salter's Brewery, and even a bakery, had their own private canal branches.

Below Rickmansworth, **Stocker's Lock** took over as the

Checkpoint Charlie for coal coming southwards in 1861. Both the toll-keeper's and lock-keeper's buildings remain standing. From here – somewhat surprisingly, given the number of conurbations we pass near to – there is some superb scenery down towards north-west London. At **Harefield** the canal crosses into the ancient county of Middlesex, past woodland as well as urban areas, and near to the pretty village of Denham. Nearby were the famous Studios, where *In Which We Serve*, *Brief Encounter* and other stiff-upper-lip British films were made, the stages closing in 1952. **Denham Deep Lock**, as its name implies, is the greatest rise on the canal, at over 11 feet. This was effected to protect the water supply of local millers. **Denham Country Park** is nearby.

We soon arrive at **Uxbridge**, which was once another important canal centre. King's Mill flour used to be made near Uxbridge Lock. The lock cottage and a turnover bridge still survive. Fellows, Morton and Clayton had a dry dock here, as well as a large boatyard, where wooden boats were built. It's now a boat centre.

Between Cowley and Southall, a bewildering number of different docks and short branches were built to serve local brickyards, from the products of which the north-western suburbs of John Betjeman's 'Metroland' were built. All are now vanished, as are their branches, long ago filled in. Only faint traces of their ever having existed remain.

At **Cowley Peachey Junction** a regular passenger service ran to Paddington – until the railways put it out of business. Here the five mile **Slough Arm**, completed in 1882, branches off into what, since 1974, has been Berkshire.

Running past the old Nestlé and EMI factories at Hayes, the canal reaches **Bulls Bridge** at Southall. Here the lock-free **Paddington Branch** begins, while that to Brentford continues south, for five miles. Bulls Bridge was once a substantial depot of the Grand Union Canal Carrying Company, where narrow boats would lay up waiting for jobs to come in. In the 1950s, under the control of the British Transport Commission, this was done by tannoy, as 'Steerer So-and-So' was invited to pick up his (or her) docket. ('Captain' by this time being an appellation reserved for a 'Number One', who owned his own boat.) Bulls Bridge can be seen at work in several historic documentary films. There was a floating school here, and even a small maternity ward for the boating families.

The journey from Paddington westwards, according to John Hassell, in his 1819 book,

is a pleasing presage of what may be expected. Having turned our backs upon the metropolis, we find at the commencement of the canal, emerging from the capital, a very beautiful burst of scenery, the Hampstead hills ranging in a picturesque curvature, crowned by wood, and ornamented with their churches and villas, passing off in a north-west direction towards Kilburn and Wilsden-green; the lofty woods of Kensington on the left, lead away to Holland-house, beyond which the country opens for a considerable space, forming another amphitheatre of pictorial beauty, until it reaches Old Oak Common, and the rising knolls beyond Holsden [Harlesden] green, where it unites with those hills that pass from the opposite side.

It is fair to say that this pleasing invocation of sylvan splendour is no longer quite so accurate a description of the Paddington Branch. Opened in 1801, it runs 13 miles to Little Venice, east of Paddington. This stretch may not now be pretty, but the Grand Junction's viability as a canal in the early twentieth century is proved by the large numbers of important factories that set up between Bulls Bridge and Old Oak Common, including J. Lyons, Glaxo, Guinness, Heinz and McVitie's Biscuits. All had their own docks or wharves, in the days when coal was still the main source of fuel. All are now gone as commercial concerns, and some buildings have disappeared altogether. The names only survive in the canal enthusiasts' lexicon. The 'Jam 'Ole' was where the old firms of Ticklers and Kearly and Tongue made confectionery. The last shipment of Black Country coal arrived here in 1970. Nearby is the steep **Horsenden Hill**, a much-used leisure amenity. The canal then passes in an aqueduct over the North Circular Road, reminding us we really are in London. Wembley Stadium, opened in 1923, is in the distance.

Past **Old Oak Common** the

canal continues by Wormwood Scrubs – not just a prison, but an area of common ground that's a haven for rare birds. Passing by disused but still rather elegant gasometers, we come to the vast **Kensal Green Cemetery**. Water gates indicate that customers once came for burial via the canal. At **Portobello Dock** a disgusting trade in refuse was based, taking London's filth to fill up the old brickyards further west, one of the dirtiest jobs a boater was asked to do.

Then comes a terrace of buildings at **West Kilburn** which, from the Harrow Road, looks typical of thousands of others. But on the other side they abut the very edge of the canal, a strange sight. Just before **Half-penny Footbridge** (so-called because that was what you paid to cross over) is **Paddington Stop** – now a pub but once a toll-house of the same name. The canal now passes through **Maida Hill Tunnel** and thence to the sudden charm of **Little Venice**.

Legend has always had it that this chic waterside centre was given its name by the poet Robert Browning, who used to live nearby. It seems more likely that it was first used in a novel by Margery Allingham. Nevertheless the small island in the pool here is still called **Browning's Island**. To the south is the large, newly developed **Paddington Basin**, once only dwarfed in importance in London by the City Road Basin, on the **Regent's Canal**, which starts here. Beginning in 1951, pleasure trips from Little Venice down to Camden were given by one of the pioneers of canal restoration, John James, in his famous boat *Jason*. It's still going.

Canal Mania

The American War of Independence had called a temporary halt to canal building, but once it finished, in 1783, trade, especially in cotton, picked up, and capital once more became available – and was itching to find a home.

The shareholders of the early canals were making good money out of their businesses. In 1789 the Birmingham Canal Company dividend stood at 23 per cent. Shares in the Coventry Canal doubled in price as soon as the Oxford Canal was completed in 1790. By 1792, fifteen years after it was finished, shares in the Trent and Mersey Canal had increased fivefold in value. The same financial success story was, broadly, repeated in other early canal companies.

That there was big money to be made in canals didn't go unnoticed. In the 1790s, canals became a honeypot for investors and speculators. This feeding frenzy became known as 'Canal Mania'.

In 1790 one Canal Act was authorized. In 1791 there were seven; in 1793, the height (or depth) of Canal Mania, twenty. Between 1791 and 1796, fifty-two new canal companies were authorized, with a combined capital of more than £7.5 million pounds. Then, an astonishing sum.

At the most febrile stage of the madness, promoting a

new canal company was pretty much a licence to print money. In just one hour the Rochdale Canal subscription book took in £60,000. Even the far more modest Grantham Canal, a rural waterway, raised the full £40,000 it was empowered to solicit in just one day. On payment of a deposit, a subscription entitled you to become a full shareholder if an Act was passed to authorize the canal.

So hot were tickets to this new lottery that people who promoted new canals even took to trying to keep subscription meetings secret (given that the number of would-be subscribers was fixed). On one occasion, in Somerset, the promoters actually bought up every copy of the newspaper in which details of where and when the public meeting was to be held had been advertised.

Manoeuvring narrow boats in the dock with barge poles
on the Grand Union Canal, 1950s

Little wonder they wanted to keep the game secret. Shares were by now changing hands for double their nominal price. Little wonder too that a bubble blew up. Or that it burst.

In part this was due to another revolution – that of 14 July 1789 – and the wars against Revolutionary France, and then Napoleon, from 1792 until 1815 (with a pause for breath in 1814). Prices rose dramatically. This, of course, meant that the money ran out more quickly than had been bargained for. Shortage of available capital then left the project floundering. What's more, the estimates for the construction costs of these canals were often woefully over-optimistic. A vortex developed.

Shareholders were required to pay up, not in one go, but in stages – 'calls' – as the money was needed. Some continued to do so and even paid more than they had first reckoned on in order to keep the project alive. Others, once it looked like a canal was doomed, took the decision to forfeit their deposits rather than throw good money after bad. The shortfall had then to be made up by further share issues, or by loans of one kind or another.

Debts could be crippling. Most of the canal companies continued paying dividends, in order to keep their shareholders onside. This diverted money that could have been used for completing the canal, thus compounding the delays in construction and putting the possibility of actually making a profit further and further into the distance.

Some of the mania-period canals were sound business propositions and, when finished, made useful and important links between the earlier navigations. In this category fall canals such as the Derby, Nottingham, Barnsley, and Wyrley and Essington.

Many short canals were built to link up with mines, quarries, iron works or other large industrial concerns. It's these stretches that sometimes look baffling when one glances at a map of the canal network. They seem to lose steam and peter out in the middle of nowhere. But that's because the business they were built to serve is no longer a going concern. Nowhere was once somewhere.

These were the success stories. Many of the canals excitedly subscribed to in this period of hysteria never got past the parliamentary hurdle. Many that were authorized were not completed until well into the next century. In some cases, not until Queen Victoria was on the throne. The Herefordshire and Gloucestershire Canal, authorized in 1791, wasn't fully open until 1845. Others, such as the Dorset and Somerset Canal, were never finished at all, but quietly abandoned and left to their fate.

Another reason many of the mania-period canals didn't make money was because of competition from existing canals, which naturally lowered their tolls. Good news, of course, for the customer. The fact that there were three trans-Pennine canals is a case in point. (The Rochdale and the Huddersfield Narrow Canals were both products of Canal Mania.)

Meanwhile, some canals that were built never had the ghost of a chance of succeeding – most notoriously, the so-called 'agricultural canals'. Because they were in less populated and less industrial parts of the country, demand was never high enough to make them viable in the first place. And when imported, cheaper grain started arriving by sea, from the 1870s, they suffered further pain. The Basingstoke Canal is an excellent example of an agricultural canal that failed.

Lastly, the new canals, in their frenzied eagerness to get their Authorizing Acts past the opposition of the existing companies, often made the mistake of agreeing to disastrous compromises.

A classic example of the debilitating effects of one such compromise is offered by the Ellesmere Canal, which was conceived as an ambitious network linking the Mersey to the Severn, running through the iron foundries and coalfields of north Wales. It didn't succeed in its ambitions. But it did see the building of one of the most celebrated pieces of engineering in Britain. If not the world.

The Llangollen Canal

Introduction by John Sergeant

Travelling on a narrow boat as it goes up and down hills through a series of locks soon becomes routine for today's boatmen. There are exceptions, when the flight of locks seems almost endless, and it takes hours to complete, but to go through a lock or two can be a very pleasant way to break up the journey. And if you have a young crew they often like nothing better than to leap off and sort out the lock mechanism.

But there is an entirely different kind of enjoyment when you travel by narrow boat across an aqueduct. You seem to be defying gravity and even common sense. The canal is not resting on the bottom of a valley or cutting straight through a field; it is carried bodily across a bridge. Seen from a boat, the pencil-like waterway flies like an arrow across the deep gorge. On the first occasion it can be quite frightening. There are no railings. You look down on either side, and there is a sheer drop. But when you reach the other side there is a thrilling sense of achievement.

Those at least were my feelings when I went across the two great aqueducts which grace the Llangollen Canal. The first one, built in 1801, crosses the valley of the River Ceiriog and takes the canal into Wales. The second, constructed four years later, is the mighty Pontcysyllte

Aqueduct, which soars across the River Dee. It is one of the most impressive feats of canal engineering in the country and helped make Thomas Telford, its designer, one of the most famous men of his time.

The aqueducts are in every sense the high points of the Llangollen Canal. But there are many other delightful aspects to this journey. I found the scenery particularly attractive. Long stretches of the canal go through woods which are largely unspoilt. When it was built, swathes of land on both sides of the waterway were bought by the canal company. Gliding through these woods today, often as the only boat to be seen, you feel as if you are being ushered through the grand park of a stately home.

The canal's original purpose was to take heavy slate and other building materials out of Wales to be exchanged in England for machinery and all the other products of the industrial revolution. One of the bulk goods from Wales was quick lime used as fertilizer on fields and in mortar on the canal system's brickwork. At Llanymynech, there is a vast empty kiln, shaped like the crypt of a cathedral, which was used to heat up the limestone rocks. I was shown how quick lime, dropped into a glass of water, bubbles furiously at high temperature, as if it is about to explode. The canal workers had to transport it carefully in its bone-dry state, keeping the rain out with tarpaulin sheets and praying their boats didn't spring a leak.

As we approach our final destination at Llangollen the canal narrows sharply, and only one boat can get through at a time. This stretch had to be modified some years ago after storm damage led to a collapse of the old banks. A full repair would have been too expensive, so this compromise was reached. I was tempted to direct traffic as canal users ran ahead, shouting into their mobile phones, letting each other know when the way was clear. It is a good example of how modern methods have allowed these waterways to stay alive. And I am sure that the eminently practical and level-headed Mr Telford would have thoroughly approved.

The Llangollen Canal, as it's called today, began life as one part of a far grander scheme, the Ellesmere Canal. But the canal was a delinquent, defaulting on its ambitions. It was never completed. But though it didn't arrive, thanks to the Pontcysyllte Aqueduct – now a UNESCO World Heritage site – it certainly travelled hopefully.

An initial meeting to discuss the building of the proposed canal was held in the Royal Oak, in Ellesmere, in 1791. (It was common to hold such a meeting somewhere around the halfway point of the canal proposed.) Ominously, this year was the start of what would spiral into the Canal Mania. The original aim was to link Netherpool, on the Cheshire bank of the Mersey estuary (its name was later changed to Ellesmere Port), to the River Dee at Chester, and then on to Shrewsbury, and the Severn, the route to run through Wrexham, Ruabon and Ellesmere itself.

However, as was often the case when a canal was first suggested, rival interests got involved, who favoured a more easterly route than the one originally proposed. The main idea here was that the Chester Canal would form part of the new Ellesmere Canal. And the Chester Canal Company needed a break. In the 1770s, fearing that the ever-growing trade into the Mersey estuary was turning the Dee estuary into a backwater, the city of Chester had spent large sums improving the navigation of the river, and a canal was built from Chester down to Middlewich. Here, however, at the instigation of the Duke of Bridge-water, the Trent and Mersey Company had refused it permission to join its canal – or indeed come within 100 yards of it. For the preceding ten years, the Chester Canal had thus been out on a lonely limb. The proposed Elles-mere Canal, if it could be forced to use part of the line, could be its salvation.

The Llangollen Canal, *c.* 1935

To decide on the best route, an 'Engineer of approved Character and Experience' was now sought. The choice fell on William Jessop, builder of the Selby Canal and already one of the most experienced of canal engineers. He was also by then heavily involved in iron work, hav-ing co-founded the Butterley Iron Works Company in

Derbyshire in 1790. The Llangollen would become famous for its pioneering use of the metal. Jessop now suggested a different route again – which was adopted.

In September 1792 a subscription meeting was held, gripped by what the chairman of the Ellesmere Company later called 'a paroxysm of commercial ardour'. Hordes of the usual suspects found their way to the venue, and the promoters did their best to defend the project 'from the excessive intrusion of too ardent speculators . . . The books were opened about noon, and 'ere sun set a million of money was confided to the care of the Committee.' Only a quarter of this was accepted.

Backed by the Chester Canal Company, the rivals still favoured a route that utilized part of that canal and sought subscriptions for their own scheme. The Ellesmere promoters compromised with the offer to build a branch to the Chester Canal from Whitchurch to Hurleston, near Nantwich. The two groups now joined together in harmony, and an Authorizing Act was passed in 1793.

The idea of a canal that would link the Mersey estuary and Chester to Shrewsbury was a perfectly sensible one. North-east Wales and Shropshire contained some of the most important iron works in the country, as well as coalfields and limestone quarries.

One of the foundries to be served, at Bersham, south-west of Wrexham, was owned by the great proselytizer of the merits of the metal – John 'Iron Mad' Wilkinson. Wilkinson was one of the pioneers of the production of cast iron. He also developed a means of blowing blast furnaces to make them hotter, using a hydraulic-powered engine.

For Wilkinson, no object was considered beyond iron's reach. He was the main force behind the famous Iron-bridge. In 1787 he had built, and sailed, the first ever iron boat. Called *The Trial*, it was launched on the Severn – at Ironbridge, naturally. 'It answers all my expectations,' he declared, 'and has convinced the unbelievers, who were 999 in a thousand.'

The River Dee near Denbighshire in north Wales, 1930s

Wilkinson died the dream as well as living it. He was buried in an iron coffin, his grave topped with a valedictory obelisk – made, of course, of iron.

It was Wilkinson and his fellow ironmasters, quarry and colliery owners who were to be served by the new canal, with branches built to their premises. Bersham would be linked, and so too the extensive iron works at Brymbo nearby.

By 1795 the Wirral Line was completed, from the Mersey

estuary at Ellesmere Port (as it would be named the following year) to Chester, where it joined the Dee. Two years later it linked to the Chester Canal.

The Wirral Line did a brisk trade in passenger transport. This stretch of the waterway was built broad so that it could be sailed by Mersey flats. Elsewhere, to save money by avoiding the construction of wide tunnels, Jessop and the committee decided to build the locks narrow.

In 1796 the Llanymynech Branch (later to become part of the Montgomeryshire Canal) was completed, running to the major junction at Frankton, just west of Ellesmere. This made a connection to the extensive limestone quarries around Llanymynech, which provided construction materials for the rest of the line, and was an important source of revenue, because of its use in iron-making and as a fertilizer.

So far so good. But then the canal started to unravel.

In 1797 a short section of the proposed link to Shrewsbury had been built, from Frankton down to the charmingly named Weston Lullingfields. But by this time the rival Shrewsbury Canal had been finished, and it became clear that its chains of distinctive 'tub boats' could carry the coal and iron of Coalbrookdale and of east Shropshire more cheaply than could the Ellesmere. The connection to Shrewsbury – for which a sizeable and thus expensive tunnel would have been needed in any case – was never finished. The canal ended high and dry at Weston.

The line that was meant to run from Chester down to Pontcysyllte, via the collieries and iron foundries of Ruabon and Wrexham, never came to be finished either. Opposition from landowners delayed it, but the main problem was, as with the Shrewsbury Line, economics.

An extensive system of tramroads had grown up in the area, and they could carry the coal and iron more cheaply than could a canal. Here was a sign of things to come.

Only the tiny Ffrwd Branch, all two miles of it, was built, and it was probably never used.

The present shape of the Llangollen Canal, with its curious dog leg, running south-west from Hurleston and then turning north-west at Frankton, is down to the persistence of the Chester Canal. They had already persuaded the Ellesmere Company to build a branch of its line to join their canal at Hurleston. When the Ellesmere dragged its feet, the Chester Company threatened to cut off the water supply for the Wirral Line, which it provided. Reluctantly, the Ellesmere Company acceded, and now the branch became the main line. The water for this was supplied from a feeder branch at the other end of the canal, built from Llangollen to Llantysilio.

The Ellesmere Canal, as it was still known, having missed out on most of the iron it hoped to carry, transported coal, limestone and other building materials within the local area. 'Beginning nowhere and ending nowhere', as L. T. C. Rolt put it, it remained a closed system, a canal apart from the national network.

Although William Jessop planned and supervised the engineering of the canal, the name most closely associated with it is that of Thomas Telford, one of the most famous of British civil engineers, who began work there in 1793. He started off as one of several engineers working under Jessop, but was to play a major part in the building of the canal's two most famous features, the Chirk and Pontcysyllte Aqueducts, of 1801 and 1805.

Born in 1757 in Dumfriesshire, Telford was apprenticed as a stonemason. He then worked on the building of Somerset House, under the architect Sir William Chambers. His big break came when the richest individual in Britain, Sir William Pulteney, employed him at Bath. Pulteney was also a big noise in Shropshire, where he was MP for Shrewsbury.

Thomas Telford (1757–1834), engineer

Through his influence Telford became the surveyor of public works for the county, and ended up building some forty bridges there (some necessitated by flooding), setting him on his way to becoming a leading civil engineer. Telford also built harbours, piers and even churches. (A glance at the latter shows that his true calling was not in architecture, but in engineering.) He later went on to become one of the most important British road builders, known as the 'Colossus of Roads'. He was also a published poet.

The Llangollen gave Telford his start in canal building. He was appointed 'General Agent, Surveyor, Engineer, Architect and Overlooker' at the age of thirty-six. His role was to help plan the canal as well as build it, and he was charged to submit 'Drawings to the Consideration and Correction of Mr William Jessop'.

In recent years there has been a great deal of controversy within canal history circles surrounding the two men's specific involvement in the Llangollen Canal. Telford has been accused of deliberately underplaying Jessop's contribution in order to boost his own reputation – even, it has been alleged, by falsifying the records concerning the Pontcysyllte Aqueduct. Whether it was Telford, as he claimed, or Jessop who came up with the idea of constructing an iron trough to cross the river is the main issue at stake.

A view of the Pontcysyllte Aqueduct in north Wales, 1806

Unquestionably, Jessop, a modest, rather unassuming individual, despite his prominence and prestige, played a vital role in the construction of the Llangollen Canal. The relative contributions of the two men can perhaps best be gleaned from the oration given at the Pontcysyllte Aqueduct's opening in 1805. It praised 'Mr Telford who, with the advice and judgment of our eminent and much respected Engineer Mr Jessop, invented, and with unabated diligence carried the whole into execution.'

In 1813 the Ellesmere Canal amalgamated with the Chester Canal, quickly deciding to call the new concern the Ellesmere and Chester Canal. It got a considerable boost to its viability when the Birmingham and Liverpool Junction Canal was opened in 1835. This joined up with the Trent and Mersey Canal at Middlewich. Finally, the Chester Canal was now allowed to trespass on to the sacred waters of the Trent and Mersey too, and so the Llangollen was linked not just to the Dee estuary but to the wider world.

The Llangollen Canal – The journey itself

St Martin's Bridge near Preesgwyn

The Llangollen Canal is one of Britain's most popular, and most beautiful. One reason for this is that it winds its way through countryside, often remote, and unspoilt by industry of any kind. The other is the famous Pontcysyllte Aqueduct. The 46-mile route is lined with attractive bridges, built with engineering bricks, and several 'bascule' drawbridges. There are 21 locks. For long stretches, hardly a building is to be seen.

Our route begins on the Chester Canal at **Hurleston Junction**, near the important salt town of **Nantwich**. The canal climbs through four locks, three with side ponds, the fourth joined to Hurleston Reservoir. They are unusual in having only one top gate, made of metal, instead of the normal wooden pair. After passing under several attractive bridges, and by a 'winding' area, where boats can turn, we come to four more similarly designed locks at **Swanley**, with its large marina, and then on through flat farmland, traditionally known for its dairy produce – not least, Cheshire cheese, Britain's oldest variety.

Approaching Wrenbury, the canal passes near to the impressive faux-Elizabethan **Wrenbury Hall** (the estate itself was founded in the fourteenth century). It is now a wedding venue. In the middle of the picturesque village, the canal crosses the River Weaver, at this point hardly more than a stream. There are several thatched cottages, and the strik-

ing red sandstone **St Margaret's Church**, which dates from the early sixteenth century. The nearby **Cotton Arms** is named after a local landowner, while at **Wrenbury Wharf** the **Dusty Miller** is a converted mill, as the name implies. Another former mill is now a boatyard.

Just past Wrenbury is the first of a series of bascule, or lift, bridges. Until recent times they were wooden and operated by means of counterbalanced beams, the design of which is unique to this stretch of the Llangollen Canal. This particular one is motorized, while others further down are still operated by hand. Steel has replaced the wood, and hydraulics now help provide the power.

The canal passes through more very rural, gentle scenery, past **Combermere Abbey**, a monastery-turned-stately home. A monument commemorates a military member of the Cotton family. Very few buildings intrude – although the A49 makes its baleful presence felt. We enter Shropshire at **Grindley Brook**. Here is a three-lock staircase, the canalside and railway bridge lined in places with distinctive blue engineering bricks. The lock-house is a good example of Telford's undoubted architectural skills.

The canal now takes a sharp left-hand U-bend to pass near the market town of **Whitchurch**. Once on the original route up to Chester, the town was instead served by a separate branch. In 1944 it was abandoned and subsequently filled in. Recently restored, in part, it now serves as moorings. After Whitchurch the canal turns south and runs parallel with the nearby Welsh–English border, through more unspoilt countryside, eventually joining the border at **Fenn's Wood**.

The wharves and warehouses at nearby **Platt Lane** are relics of the time when the canal would have been dotted with such small commercial centres. The **Waggoners Inn**, nearby, indicates the fact that this was once a hub of transport as well as a hive of industry.

At **Whixall Moss** is an attractive cast-iron roving bridge. Here the **Prees Branch** spins off from the main line, built to connect to a series of limestone quarries and kilns. On it are still to be found two lift bridges made of wood. A mile of the original four miles is in use.

To the north and west is an area of peat bog, one of the largest in the country and now a National Nature Reserve. During the Second World War, the military used to set fire to large swathes of moss at Fenn, Whixhall and Bettisfield, hoping to persuade Luftwaffe pilots that they were bombing Liverpool.

The canal now heads towards Bettisfield, briefly straying into Flintshire, Wales, before returning to Shropshire to pass by a series of very pretty meres, or lakes, surrounded by woodland and teeming with birdlife. Before the final one – **The Mere** – we pass through the **Ellesmere Tunnel**. Ellesmere, today a pleasant market town, of course gives the canal its original name. By the canal is **Beech House**, the company's former HQ, designed by Telford. A weather vane on the roof of the old dry dock is made in the shape of a narrow boat. Until 1961 lock gates were made here. A branch runs into the town itself. At **Ellesmere Wharf** are several old canal buildings and a crane that survives from the heyday of the once-busy basin.

The waterway now winds through hills and farmland on to **Frankton Junction** – a small hamlet, but the hub of the whole canal. Here the **Llanymynech Branch** runs south, through a series of pretty locks, to link up with the Montgomery Canal, which joins up with the upper reach of the Severn at Welshpool, near to where the river rises. (Originally it was known as the Montgomeryshire Canal, as it does not actually go to Montgomery. Though built by the Ellesmere Company, this branch became a part of the Montgomery Canal.) A short way down this line, the **Weston Branch** heads off for Weston Lullingfields, in its forlorn pursuit of Shrewsbury.

From Frankton the main line now turns north-west, through flat farmland, studded with woods. The last two locks are at **New Marton**. The land starts to become hillier as we approach **Preesgwyn**, a former colliery town, part of the Denbighshire coalfield. Towards Chirk and the Welsh border the spectacular scenery that gives the Llangollen Canal its renown unfolds.

The emphasis on the extraordinary starts with the second most famous of the canal's mighty constructions, the **Chirk Aqueduct**. The River Ceiriog, which it crosses, is the border between England and Wales. Engineered by Telford, completed in 1801, it carries the waterway – and towpath – 70 feet over the river. The original plan had been for a series of locks, but Jessop proposed a towering aqueduct, which, he said, 'instead of an obstruction, would be a romantic feature of the view'. He was right, although for some travellers the fact that the railway line right next to it is taller now rather spoils the effect.

The sides of the aqueduct are made of stone. The water is carried, though, on iron plates. This meant that the structure was lighter than it would have been had it been built using the old method of masonry treated with puddled clay. Thus putting less pressure on the stonework. Later, a complete iron trough was fitted within the stone sides. The fact that Chirk was something of a halfway house is down to the fact that a complete iron trough, *sans* masonry, was unprecedented, and the engineers seem to have got cold feet and hedged their bets.

Soon after crossing the river the canal enters the 459-yard **Chirk Tunnel**. Unusually for its time, it was built with a towpath, made of stone. Telford had long considered legging a dangerous, undignified and demeaning occupation for the boatmen. To the west of the tunnel is **Chirk Castle**, first built in 1295, surrounded by gardens, parkland and thick woods.

The Chirk Tunnel is followed by **Whitehouses Tunnel**, also fitted with a towpath. On top of it is the Holyhead Road – built by Telford himself. Now the A5, it runs from London to the Menai Strait, shadowing the old Roman road of Watling Street, before crossing to Anglesey over Telford's famous Menai Suspension Bridge. From the end of the tunnel the scenery is spectacular, with the Welsh mountains in the distance.

At **Irish Bridge** there's a signpost for Offa's Dyke, the eighth-century fortification built by King Offa of Mercia to protect his borders from attacks by the Welsh. Its route is roughly that of the border between Wales and England today.

Then comes another master-piece of construction: the **Pontcysyllte Aqueduct**, completed in 1805. (Pontcysyllte means 'the bridge that connects'.) Again, it was initially proposed as a conventional masonry-plus-puddling construction, but it was Telford (or so he claimed at the time) who decided to build the aqueduct as an iron trough, duly submitting his plans for the approval of Jessop. By this stage Telford had just completed the building of the world's first ever iron trough aqueduct, at Longdon on the Shrewsbury Canal. Although that was a pygmy compared to the project now in hand, it proved that the concept worked.

Telford's original conception was for there to be seven 60-foot spans between the piers. Jessop suggested eight would be safer, giving more support. The size of the cross-section of the columns carrying the trough was also increased, at Jessop's suggestion, by just over a foot. Jessop feared that 'I see men giddy and terrified in laying stones with such an immense depth beneath them.'

Finally, it was decided to extend the aqueduct on the south side, rather than build an enormous embankment to meet it. Hence its current appearance, with eighteen arches of 53 feet each. The resulting embankment, a massive 97 feet high, is a considerable feat of engineering in itself.

The nineteen stone piers were built hollow, with cross walls inside them, rather than following the usual method of filling them with rubble (which, as Telford remarked, added weight but not strength). This allowed the arches to be more slender, giving them their elegant, graceful appearance. Rumour had it that the blood of young bullocks was used to make the mortar for the brickwork, the joints of which were then rendered with Welsh flannel, boiled in syrup.

The water is carried over in a cast iron trough, 1,007 feet long, with a towpath and handrail on one side and a sheer vertiginous drop on the other. At its highest point it is 127 feet above the foaming River Dee. The towpath is fixed above the water, resting on cast-iron pillars. The trough was made at the important Plas Kynaston Foundry, within view of the aqueduct itself. (A small branch line ran to the foundry.)

On seeing Pontcysyllte, the future Poet Laureate, Robert Southey, was lost for words:

Telford who o'er the vale of Cambrian Dee
Aloft in air at giddy height upborne
Carried with navigable road . . .

While the question of precisely how much William Jessop added to the design of the construction will probably never be solved, one piece of evidence that Telford, in order to secure for himself the reputation of *Pontifex Maximus* (as he was indeed known), kicked over the traces of Jessop's contribution is to be found on a plaque on the base of one of the piers. It reads:

Thomas Telford was Engineer
Matthew Davidson was Superintendent of the Work
John Simpson and John Wilson executed he Masonry
William Hazledine executed the Iron Work
William Davies made the Earthen embankment.

One name seems to have been left off the roster.

At Pontcysyllte is a pretty, tuning-fork-shaped canal basin

that reaches towards Trevor. This is a stub of the original main line that was meant to continue to Ruabon, and up to Chester.

The canal now turns sharply south-west, running along the steep sides of the valley, with many fine views of woods and limestone cliffs, until it reaches **Llangollen** itself, on the River Dee. This pretty town was a very popular inland resort in Victorian and Edwardian times. A canal museum is located in a former warehouse on **Llangollen Wharf**. From here horses once more tow narrow boats, now full of sightseers, along the final section of the canal.

The route comes to an end at **Llantysilio**, overlooked by the Berwyn Mountains, next to the distinctively shaped weir built by Telford to bring water to the canal at **Horseshoe Falls**. Hugh McKnight, in the *Shell Book of Inland Waterways*, calls the last six miles of the Llangollen Canal 'the loveliest in Britain'.

The Pace Quickens

By the early 1790s British overseas trade was booming. Canals brought the finished goods to the ports, but the arrival of the boats was not something you could set your time-piece by. And delays could be disastrous for the merchants concerned, who incurred expensive ware-house charges, sometimes for several months at a time if the ship was missed, or lost their window of opportunity in their foreign market altogether. 'Want of certainty' meant that high-value or perishable goods still travelled by roads, as the canals were not efficient enough to guar-antee delivery.

The pressure increased on the canals and carrying com-panies to rectify this state of affairs – especially as competition between them was now hotting up. The result was the introduction in the 1790s of fly-, or fast, boats, which ran to a timetable. On payment of a higher fee to the canal, fly-boats worked around the clock, manned by two teams of two men working in shifts, with fresh horses at each stage. These crack outfits could cover around 40 miles a day. A working 'day' boat would expect to travel around 25 miles per day.

The fly-boat usually carried around 15 tons maximum, made up of higher-value goods. A narrow boat would carry around double that weight. Bulkier cargo, not subject to the same time pressure, remained on the day boats.

Pickfords

One early fly-boat company was Pickfords, still famous today for their removal vans. Their history encapsulates that of British transport since the late seventeenth century.

In 1646 we find a Thomas Pickford running a family business engaged in mending roads. To obtain stone he made regular trips to a quarry he owned, south of Manchester, using packhorses. Then he had an idea. He began to make sure he made the return journey laden with another cargo. He had become a 'carrier', or carter.

A Pickfords boat beneath Macclesfield Bridge, Regent's Park, 1827

By 1720 the business was doing well enough for James Pickford to have his main offices in the City of London, operating a wagon business between London and Manchester. Fifty years later the family invented the fly-wagon (precursor of the fly-boat) which was capable of carrying goods or passengers from London to Manchester in the then astonishing time of four and a half days.

In 1778 the far-sighted Pickford family dipped their toes into the canal carrying business, acquiring twenty-eight boats the following year. Once the canal network in London was extended Pickfords wharves were built in the City, Paddington and Brentford Basins, and they owned other wharves and depots along the routes to Birmingham, Manchester and Liverpool. They became the most extensive and important of the independent carrying companies.

In 1840, ahead of the game as usual, Pickfords pulled the plug on their canal carrying business and concentrated exclusively on railways and roads. By the turn of the next century the firm was heavily invested in road haulage. Once again they had seen the future. Then they reinvented themselves again, as the removals company we know today.

The Good Times

Once the Napoleonic Wars were over in 1815, after a brief slump, canals thrived. Goods for export were sent by the manufacturer to the port with the best shipping connections to the end market – Liverpool for the West Indies, Hull for Russia and the Baltic. This meant that often the inland water journey was a long one. The biggest market for finished goods, and the biggest port, was London, and so the fly-boats often came to and from the capital. Large wharves and warehouses were built along the Regent's Canal, in particular.

Most trade on the canals came from Birmingham, but there was also cheese from Cheshire, pottery from the Potteries, woollens from Yorkshire, cutlery from Sheffield, cotton goods from Manchester and lace from Nottingham. The total mileage of Britain's canals was still growing dramatically. In 1760 there were 1,398 miles of canals and navigations. By 1800 this had more than doubled, to 3,074. In 1820 the total stood at 3,691.

In 1793 100 boats a day passed over the Birmingham Canal summit. In 1821 it was more like 400.

A wooden bridge over the Llangollen, which could be raised when boats needed to pass

The canal company shareholders were raking it in. In 1810 the Mersey and Trent Company paid a dividend of 40 per cent, which would nearly double by 1821. Elsewhere profits were flooding in. In 1833 the Oxford and Coventry Canals were paying 32 per cent and the Erewash Canal 47 per cent, dwarfed by the Loughborough Navigation, which in 1827 was yielding in excess of 150 per cent. Pickfords and the other large independent carriers were coining it too.

What could possibly go wrong?

The Duke of Bridgewater had thought of something. Not long before he died, in 1803, watching a string of coal barges passing along his canal he remarked, 'Well, they will last my time, but I see mischief in those damned tramroads.' In the event the canals' heyday lasted nearly forty years longer than did His Grace. But the writing had been on the wall for the whole of that time.

The Coming of the Railways

Tramroads, wagonways or tramways are an oft-forgotten but hugely important means of transport in British history. They began in England in the early seventeenth century, an idea copied from Germany. They were found particularly in the areas associated with coal and iron – Shropshire, Newcastle, south and north Wales (as the Ellesmere Canal had found to its cost).

Tramroads were initially built of wood. But from the 1760s rails began to be tipped with iron strips. In 1767, at Coalbrookdale, the Darbys produced cast-iron rails.

Benjamin Outram then popularized the use of stone sleepers and introduced L-shaped, flanged rails that were used in conjunction with wheels that were flat. This was superseded by a system designed by his partner, William Jessop, in 1798, in which the wheel was flanged rather than the rail. This was the design followed by subsequent railways. In 1820 wrought-iron rails, around 15 or 18 feet long, were introduced by John Birkenshaw, in Northumberland, and these became the standard.

The tramroads of the hilly terrain of the south Wales iron and coal industries were on inclines. Gravity pulled the wagons down the track, while a horse (sometimes men) pulled them back up. (There was one extraordinary system whereby the wagons were propelled by sails.) More cleverly, systems of counterbalanced wagons were put in place, so that the descending laden carts pulled the empty ones back up. When canals arrived in south Wales they didn't replace the tramroads, but they ran alongside them. With their many locks, they could never have coped with the vast amount of coal and iron that needed to be transported.

Tramroads played a crucial role elsewhere in the network. When canal companies ran out of money before a canal was complete, as they regularly did, a tramroad was often used to connect the two sections, as at the Lancaster and Ashby Canals, where the tramroad was permanent, or the Blisworth Tunnel section of the Grand Junction, where it was temporary.

It's sometimes been said that canal companies were rather short-sighted and naive in that they actively assisted in the building of tramroad networks. But to blame them for promoting a rival form of transport is to ignore the fact that the tramroads were thriving in areas where no canal had penetrated.

The most extensive use of them was in the north-east of England, taking coal from the collieries down to the ports of the Tyne and Wear. Here canals had never taken root. A vast network of tramroads – and at the risk of destroying suspense we may as well start referring to them as 'railways' – was in place. All they lacked to allow them to scale up and break the bounds of their local area was a locomotive engine. And that was already on the starting blocks.

Curiously, James Watt had resisted the idea of a steam engine being used for locomotion. What's odder is that his own foreman and protégé, William Murdoch, had long been urging it.

Murdoch was another of the great Scottish engineers. He was born in 1754, in Ayrshire. At the age of twenty-three, he walked down to Birmingham to ask James Watt for a job. Once there, he anglicized the spelling of his surname to Murdock.

In 1784 Murdock suggested that the company build a steam-powered locomotive, but Watt – perhaps put off by the great quantities of fuel such a device would eat up – did his best to put him off the idea. (While Boulton urged Watt, secretly, to take out a patent for the very thing Murdock had proposed.) Murdock carried on working on the locomotive engine, however, and built several models. (A story, sadly perhaps apocryphal, is that an early proto-type of Murdock's locomotive was taken for a spin and terrified a local vicar, who took the steam-belching con-traption to be the very Devil himself.)

According to the later evidence of his son, Murdock showed his model to a neighbour at Redruth – a Cornish mining engineer and inventor, Richard Trevithick. He and another engineer were, in the late 1790s, working on the development of high-pressure steam engines. Watt had also resisted high-pressure steam, perhaps believing that the boiler plates then in existence would not be strong enough to support it. But as the industrial revolution advanced, so did the manufacture of better boiler plates. The use of high-pressure steam would allow engines to become smaller, a crucial development when applied to transport.

Nevertheless, the litigious Watt promptly claimed that the new engine violated his copyright. But by the end of the century his patent finally ran out, and the coast was clear for the next step to be taken without fear of infringement. Trevithick himself built the world's first full-scale, functioning steam locomotive, which alas was never given a name. It made its maiden voyage by road, carrying a complement of excited passengers, on Christmas Eve, 1801. Even more portentously, a subsequent Trevithick engine made an outing in 1804, along a tramway at the Penydarren iron works at Merthyr Tydfil.

Meanwhile recognizable railways had been taking shape. In 1798 the horse-drawn Lake Lock Rail Road linked coal mines to the Aire and Calder Navigation. The Surrey Iron Railway Company ran between West Croydon and Wandsworth, carrying industrial raw materials and agricultural goods, from 1802. It was a whopping 4 feet 2 inches wide and had been engineered by William Jessop and Benjamin Outram. It too was run on horse-power.

(Ironically, it lost business when the Croydon Canal was opened in 1809 – a reversal of what would happen in the future.) The world's first passenger-carrying railway was introduced in Swansea – the Oystermouth Railway, also horse-drawn, which opened in 1807.

But the future clearly belonged, as did so much else in the nineteenth century, to steam. In 1812 the twin-cylinder *Salamanca*, built by John Blenkinsop and Matthew Murray, became the first commercially successful steam locomotive. It operated on the Middleton Railway, near Leeds.

Salamanca was followed the next year by the famous *Puffing Billy*, built by William Hedley and Timothy Hackworth for a Northumberland colliery, at Wylam. The oldest surviving steam locomotive in the world, it's still to be seen at the Science Museum in London.

So by 1813 the stage was set for the next great engineering genius to make his mark on the industrial revolution. He was the son of a Wylam colliery fireman and only learned to read and write, with great difficulty, in his late teens. George Stephenson began his professional life as a cowherd and then graduated to driving the horses which pulled up the coal from the mines. Here he encountered Watt's steam engines.

Just as James Brindley had, before the job description of canal engineer became a reality, served the perfect apprenticeship to become one, so too Stephenson was able to pick up an all-round education in everything he needed to learn to become a pioneer of the Steam Age. And he was, undeniably, an engineer of genius as well as being, again like Brindley, a man of much vision. His first steam locomotive, *Blücher* (named after the Prussian gen-

11. (*left*) Boats ascending
the Bingley Five-Rise
staircase.
12. (*below*) Watering hole at
the top of the staircase.

13. (*opposite*) A boat moored at Woodlesford. 14. (*top*) Engine Arm Aqueduct, Smethwick, on the Birmingham Canal Navigations New Main Line.
15. (*above*) Woodlesford Lock on the Aire and Calder canal.

16. Boats moored on the Grand Union Canal. 17. (*opposite*) A lock on the Grand Union Canal.

18. The Grand Union passing through a green clearing. 19. (*opposite, top*) Corpach, on the Caledonian Canal. 20. (*opposite, bottom*) Black Country Living Museum, Dudley.

21. Moorings at Milton Keynes.

eral, Britain's ally in the Napoleonic Wars), was built in 1814. By 1822 he had built the first railway designed to be operated without the help of animals, at the Hetton Colliery. For this line Stephenson built five locomotives.

'Puffing Billy' – the pioneer locomotive constructed in 1813 by William Hedley

In 1821 the colliery owners of Darlington secured an Act of Parliament authorizing them to build a steam railway, and Stephenson was made its engineer the following year. The new railway would carry coal from Darlington to Stockton on Tees, and thence to the North Sea. On 27 September 1825, the new line, 25 miles long, was opened. As with the Bridgewater Canal's crossing of the River Irwell more than sixty years earlier, great excitement accompanied its unveiling, and somewhere around 500 passengers crowded on to open carriages to sample the new transport. Starting off at a sedate 4 mph, as it descended towards Stockton it reached the then unimaginable speed of 15 mph.

As with the Bridgewater Canal, the railway also had a fast economic effect – in only two years the price of coal in Stockton had dropped from 18 to 8 shillings per ton.

What should have been extremely worrying for any canal company was that several canals had been proposed in the area – not least by John Rennie. They proved too expensive to build. When, in 1823, an engineer was called in to estimate the cost of completion of a canal from the Solway Firth to Newcastle, he came up with a figure of £888,000. When asked to assess the cost of a railway doing the same job, the figure came down to £252,488.

Along with their many inherent advantages – not least that they could move ten times more quickly than could canal boats – the railways were also to benefit from the great advances in canal engineering that had been made since Brindley's day, involving ever longer and wider tunnels, deeper cuttings and bigger embankments. No canal was on a larger scale than the Caledonian.

The Caledonian Canal

Introduction by John Sergeant

Canals which alter the map by cutting through a large swathe of territory are in a class of their own; and in Britain that class is headed by the Caledonian Canal, which links the North Sea with the Western Isles. When it was built 200 years ago one of the primary aims was to enable the Royal Navy to avoid the stormy seas around Cape Wrath and confuse our enemies by slipping along a natural fault line, which runs like a scar across the face of Scotland.

The canal also has a special place in the history of the British fishing industry. When herring were plentiful vast numbers of trawlers were devoted to scooping up the 'silver darlings'. Harbours in many parts of the country depended on the fishing industry. Around the Scottish coast herring followed a pattern. At the start of the year they would be found in abundance on the west coast, and the shoals would move round to the north-east as the autumn approached. The Caledonian Canal provided the perfect short cut, allowing trawlers to move quickly from coast to coast and so keep their catches at a high level all the year round.

I travelled the full length of the canal, from Inverness to Fort William, on an old wooden-hulled trawler built about forty years ago, which had been used for fishing in the North Sea. It was a sturdy craft with plenty of space below for cabins and a spacious area on the main deck where we ate our meals. It had a 400 horse-power engine, which in the capable hands of our captain kept us firmly on track. I could not help wondering what the canal folk of the past would make of an engine with the power of 400 horses. I certainly could not have managed the trawler on my own, but with two other members of the crew, we had a strong team on board.

Traffic along the Caledonian Canal is quite different from the gently ambling narrow boats which dominate most of Britain's inland waterways. The boats here tend to be much wider, and the canal much bigger. You can still come across small craft and pleasure boats, but mostly they are sea-going vessels, trawlers and yachts, which often carry sailors ready to tackle the difficulties of the open sea. I encountered skippers from the Scandinavian countries and hardy men and women from Holland and France, who seemed to think nothing of a six-week cruise round Britain with the Caledonian Canal providing a useful, relaxing break.

Nowadays it is the spectacular scenery and the sharp contrasts along the route which make this trip so interesting and exciting. Fields and gently sloping hills provide the foreground to ranges of rugged mountains, which even in high summer have snow on their summits. Ben Nevis, the highest mountain in Britain, is a looming presence over much of the western end, and along the route of the canal a string of lochs, wide and bold, make you believe you are travelling across an inland sea.

This greatly increased my enjoyment. Not only did we pass through Loch Ness, we loitered in Loch Oich and fished, without success, in Loch Lochy. In the canal parts of the trip there was plenty to see. Few details were overlooked by the builders. The original cottages for the canal workers have long, deep roofs and heavy stone facing; the metalwork on the canal fittings glistens with fresh black paint. This canal journey is an eye-opener, which cannot fail to impress. It is one of my all-time favourites.

Three great Scottish engineers would be involved in the building of the canals in their native country. Between 1766 and 1774 James Watt, while he was waiting for advances to be made in iron-making technology that would make his stream engine a reality, surveyed routes for what would become the Forth and Clyde, Monkland, Crinan, and Caledonian Canals.

John Rennie engineered the Crinan Canal, starting in 1794, while Thomas Telford was largely responsible for the construction of the Caledonian Canal.

Unusually, the government too was involved in these Scottish canals. The reason for this dates back to the Jacobite Rebellion of 1745. The Jacobites were supporters of the Roman Catholic, and Scottish, Stuart dynasty. They had never accepted the Glorious Revolution of 1688, in which James II was deposed in favour of the Protestant William and Mary (James's eldest daughter). The accession of the Hanoverian Protestant George I in 1714 was followed by a Jacobite Rising the next year. It was suppressed, but many people still supported the king 'over the water' – and not just Scots.

The Rising of '45 was a serious threat, backed by Louis XV, although the ships he sent in support were turned away by unfavourable weather conditions in the English Channel (unfavourable for the French, that is). Under the banner of Bonnie Prince Charlie, grandson of James II, the Jacobite forces got as far south as Derby.

The rebellion was ended by the Battle of Culloden in 1746. Here the Duke of Cumberland (the youngest son of George II) brutally crushed the Jacobites. Fêted back in London, he became known to the Highlanders and their sympathizers as the Butcher of Culloden. After the rising was put down, partly out of revenge, partly because they still offered a military threat, there was a determined effort to suppress the clans. A full-scale assault on Highland culture followed, with draconian measures put in place banning the carrying of weapons, wearing tartan and even playing the bagpipes (or so legend has it, if not hard fact). In 1752, after the passing of the Annexing Act, the lands and property of the defeated Jacobites were seized by the Crown. The money raised in rent was used to fund the building of roads and other amenities in the Highlands.

The Battle of Culloden, 1746

The clan leaders who remained became landowners and *rentiers* rather than warrior chiefs who financially supported their clan members. They were joined by new landlords from the Scottish Lowlands, and England, who took over the seized lands. Subsistence farming gave way to commercial farming, with the introduction of large numbers of cattle by the landowners, followed by even larger numbers of sheep. (To accommodate the new arrivals, small-holders were turfed off their traditional lands in the 'Highland Clearances'.)

With employment becoming increasingly hard to come by, in the course of half a century upwards of 20,000 Highland Scots emigrated, to North America, the Lowlands and England. Many men who stayed, meanwhile, began to serve in the newly formed Highland regiments, which were to become a crucial source of military manpower for the Crown.

It was concern about the rate of emigration from their native lands that led, somewhat ironically, to a group of wealthy and influential Scots forming the 'Highland Society of London' in the 1770s. The job-creation scheme they particularly favoured was the boosting of the fishing industry, especially that of the herring. (James Watt, in 1773, had been asked to assess a possible route from Fort William to Inverness with the fisheries in mind.)

In 1784 a visionary amongst their number, a wealthy London-based bookseller called John Knox, pressed a plan for three canals, the routes being those eventually adopted by the Caledonian and Crinan Canals, and the Forth and Clyde Canal, work on which had already started. This network – which others had also called for – would open up 'a circumnavigation within the heart of the kingdom to the unspeakable benefit of commerce and the fisheries'. The British Fisheries Society was set up with the same goals in mind, in 1786. (One of its leading lights was Sir William Pulteney, Thomas Telford's benefactor.)

Two years later the society contracted Telford to build the harbour (and town) of Ullapool, on the west coast of Scotland, and subsequently survey more potential harbours and piers in the region.

Emigration, however, continued apace. But by now the government was finally waking up to the social and economic dangers posed by this exodus. Determined to maintain the source of recruits for its Highland regiments, it decided to act.

In 1784 the seized Jacobite estates were sold, sometimes back to their original owners, and it was from this fund that government loans were made to help complete the Forth and Clyde and Crinan Canals, both of which had stalled. Up until now, the government had steered clear of involving itself in canals, taking the view that their funding was a matter for private capital.

Now they were to join the canal business themselves. In 1801 Thomas Telford arrived in the western Highlands, funded by the Treasury, to make a report on employment possibilities for the area. He recommended an ambitious programme of construction – of roads, bridges and harbours. He also suggested the building of the Caledonian Canal.

It had been a long time coming. Not only had James Watt surveyed the same route across the Great Glen back in 1773. John Rennie surveyed it all over again twenty years later. But this time the idea was taken up. In 1803, the Caledonian Canal, backed entirely by Treasury money, received its authorization.

The entrance to the Caledonian Canal with Ben Nevis in the background, 1836

Telford was appointed as principal engineer, with William Jessop as his boss, the consultant engineer. (Telford himself had told the canal's commissioners that he wanted 'the assistance and advice of Mr Jessop'.) Jessop was, as he had done at Llangollen, to play a vital role in the construction of the Caledonian Canal, until he retired from the project in 1812, owing to poor health. He was the man in charge of the strategic decisions. But the practical construction and detailed design of the canal – though it had to be approved by Jessop at every stage – was in Telford's hands.

The government was willing to fund the Caledonian Canal for five main reasons. Its construction would, of course, boost employment. It would also help the fishing industry. A canal would allow the herring fleet to cut across from the west to the east coast and lie in wait for the 'silver darlings' as they battled their way around the north coast.

The canal would also offer huge benefits to merchant shipping. Sailing around the north of Scotland was dangerous, and often time-consuming. The vagaries of the weather meant that the passage was often delayed, sometimes for months on end. With a coast-to-coast canal, it was hoped, the problem could be solved.

Both Telford and his government employers envisaged the Caledonian being part of a network of canals. The Crinan Canal, cutting through the Mull of Kintyre, had turned an 85-mile sea journey into one of only 9 miles. It thus allowed shipping an easy route from the Clyde to Fort William.

The main reason for the government's enthusiasm for the Caledonian Canal, however, was the war with France, which had begun in 1792. Both sides were trying to blockade the other's trade. In this the Royal Navy was to take the upper hand. But the French 'privateers' – independent operators licensed by their government to attack foreign ships during time of war – were a menace. This was why the prime minister, William Pitt (the Younger), was so keen on the Caledonian Canal. Not only would it allow fishing vessels and merchant shipping to travel from coast to coast. It would be built to such a great size that navy frigates could use it to escape the attentions of the enemy.

The locks were to be 162 feet long, 38 feet wide and 20 feet deep. The depth was to be achieved by making cuttings of 15 feet, and banking up the extra 5. The breadth of the pounds was to be 110 feet. In comparison with other British canals, this was a giant in the making. Twenty-five locks were originally planned, and an additional three were subsequently called for. (The twenty-ninth lock we know today was built in 1843.)

Certainly the proposed canal had several good cards in its hand. It could take advantage of a 60-mile strip of land which, despite passing through one of the most mountainous areas of Britain, with the north-west Highlands on one side and the Grampians on the other, never rises more than 60 feet above sea level. Heavy rainfall provided plenty of water, the lack of which was the main difficulty suffered by most other canals. Of the route from coast to coast, nearly two-thirds of the waterway was already in place. Loch Ness (including little Loch Dochfour) is 22 miles long, Loch Oich 4, and Loch Lochy 10.

An author in the 1840s, remarking on these favourable circumstances, declared that 'It must have been difficult to escape the conclusion that Nature had irresistibly invited the hand of man to the completion of such an undertaking.' This gentleman – clearly no conservationist – had obviously not been intimately involved in the Caledonian Canal's construction.

For one thing these handily placed lochs needed to be made navigable. Telford's estimate in 1803 was £350,000.

Jessop, rejecting the idea of turf-sided locks in favour of ones lined with stone, revised it up to £474,000. Neither estimate included the cost of land purchases, which Telford reckoned would be a mere £15,000.

Because engineering work on this scale had never been done in the Highlands (except by the military) there was little in the way of available equipment, and wagons, wheelbarrows, even tools, had to be fashioned by smiths and carpenters on site.

Iron tramways needed to be built, and wooden gates were needed for the locks. Iron for the west end of the canal came up in ships from the Dee, for the east end from the Humber. Stone quarries and limestone kilns were put to work. The workmen themselves needed to be fed, watered and, when necessary, whiskied. This was a gargantuan operation.

Work on the canal was in part mechanized, with steam-powered dredgers employed at Loch Oich (which needed to be made deeper, as did the River Ness), with further steam engines used for pumping. The engines were, of course, supplied by Boulton and Watt, the parts brought in by boat and assembled on site.

There were difficulties from the outset. As was often the case with canals, unforeseen engineering problems emerged once construction started in earnest, causing delays and extra expense. Work wasn't carried out during the winters and had to stop when the annual budget had been spent. There were regular delays when the labour force was tempted away by seasonal work, such as peat-cutting and harvesting.

The Caledonian Canal, *c.* 1900

But it was the enormous price rises of the war years that really put a spanner in the works. By 1813 Telford estimated that wages had doubled, or even tripled, since his initial budgeting, as had the price of food. The cost of timber needed to line the lock gates had also risen three-fold, thanks to Napoleon's blockade of the Baltic.

Because of the rise in prices and the delays, by 1813, though only 11 miles were left to build between Laggan and Fort Augustus, the canal was wildly over budget. Practically all of Jessop's £474,000 estimate had been spent. Telford reported to the commissioners that the canal would be finished in 1817 – but only if another £235,000 was coughed up.

But he now ran into another major snag, in the form of

MacDonnell of Clanronald and Glengarry, who treated Loch Oich as his private fishing ground. He, Cameron of Lochiel and other landowners (one of whom claimed that the locks would frighten the fish and put them off their breeding) now demanded large sums in compensation for the canal to pass through their estates. Work was further delayed as negotiations took place. (Telford, back in 1803, had made the assumption that 50 per cent of landowners, in a display of public-spirited altruism, wouldn't charge anything at all for passage through their lands. He had probably been lulled into a false sense of security because he had been dealing with the members of the Highland Society. MacDonnell alone was able to extort over £10,000, the whole bill coming to over three times the £15,000 Telford had counted on.)

The military authorities at Fort Augustus were also jealous of their property, and caused problems with the siting of the locks.

Time and money being short, and with huge pressure building up behind him either to finish the canal or stop work on it altogether, the usually meticulous Telford was now tempted to cut corners. He began to accept the use of poorer-quality stone and relied on the hope that, even though the middle section of the canal was nowhere near a minimum depth of 20 feet, dredging would in time put things right. (It didn't.)

By 1822, when the canal was finally opened for business, the tab was more than £900,000.

Its first customers sailed through from Inverness to Fort William in October the same year. Nineteen years in the making, twelve years over schedule, it had been a long haul. Now the *Inverness Courier* took the opportunity to chide those who had shown little faith. 'The doubters, the grumblers, the sneerers,' it noted with satisfaction, 'were all put to silence.' But revenge is a dish best eaten cold. The doubting Thomases – they may even have included Telford himself – were not all wrong.

Although by 1825 the canal had been dredged to a minimum depth of around 15 feet (5 feet shallover than originally stipulated), subsequent problems with leakages meant that in places it fell to 12 or 13 feet.

This, of course, severely restricted the traffic on the canal, with the larger commercial coastal vessels that were meant to use it being of too great a draught. The same was true of frigates. But in any case, the Napoleonic Wars were now long over. After the Battle of Trafalgar in 1805, Britannia ruled the waves, making the need for British naval ships to skulk out of sight along an inland waterway unnecessary.

What's more, the era of the steamship had already begun (Telford himself travelled in one when he came to inspect the canal in 1828), and they were far bigger than the craft envisaged when the Caledonian was started. The canal, even though super-sized at the time of its inception, was already proving to be too small. (And the wash of ships capable of travelling at 10 mph would damage the banks.)

As steam began to take over from sail, ships were better able to negotiate the tricky passage around the north coast of Scotland. One of the main reasons for building the canal was now a dead letter.

Fishing boats and local traffic became the main users.

Even here, there were problems, with sailing boats finding the lochs extremely difficult to negotiate. They had a worrying tendency to act as a wind tunnel. And Telford, in his desperation to save money, had opted not to build a towpath at Loch Oich.

The Caledonian Canal was never a commercial success. By far the largest volume of traffic was passenger boats, followed by vessels carrying the usual cargoes of coal, slate and timber – along with fishing boats. But in the 1830s, in a classic case of over-fishing, the herring trade began to decline.

By 1839 the total traffic on the canal was only 2½ per cent of that rounding the north of Scotland, which was now aided by better charts. Serious thought was given to closing it down. But, thankfully for later generations, it was decided to keep it going. In 1843 the canal was improved, deepened to 17 feet and provided with steam tugs to help the sailing ships. Cargoes of 500 tons could now be accommodated. But all this cost a further £228,000.

By the 1850s, the cost of maintenance was exceeding the value of receipts. The canal was running at a loss. The amount of tonnage carried further declined, and by 1905 income from passenger steamers was 20 per cent of total takings

In the First World War, though, one of the canal's original functions – to act as a strategic part of naval operations – was finally realized. In 1918 a huge shipment of mines was unloaded from American vessels at Fort William and sent across the Caledonian to Muirhead, near Inverness, for deployment in the North Sea – now, of course, aimed at a German rather than French enemy.

In the Second World War the canal also played a part in the war effort. After this, it was used only by a few fishing boats. Like all the other inland waterways, its future would lie in pleasure boating. People enjoying a journey down the Caledonian Canal today have a distinguished forebear: Queen Victoria herself took a trip in 1873.

The Caledonian Canal –
The Journey Itself

Lock at Fort Augustus

There can be little doubt that as far as dramatic and beautiful scenery goes, the Caledonian Canal offers as much, if not more, than any British canal, as it makes its way through the **Great Glen**. At its eastern end, it begins at **Beauly Firth**, a small inlet of the vast Moray Firth, to the west of the handsome capital of the Highlands, Inverness. (Culloden lies to the east of the city.) This titan amongst canals starts as it means to continue – spectacularly. The mighty sea lock at **Clachnaharry** – once a tiny fishing village, now a suburb of Inverness – is the first construction on the canal proper. And it's an extraordinarily impressive one.

To enable coastal vessels to enter, Telford built it 400 yards into the sea of the Beauly Firth, over mud more than 50 feet deep. To achieve this feat of engineering, a clay embankment was built, which, by its own weight, forced out the mud beneath it. Next, a piled 'coffer dam' was put into place, a structure that allowed water to be pumped out so the stone walls of the great sea lock could be built. The gates to the lock are made of Welsh oak.

Because of worries that sea-going vessels would ram the gates, a chain was placed in front of the entrance and only lowered once the lock-keeper was satisfied that the sailors were in control of the boat. Not, perhaps, always the case.

Enormous four-arm capstans opened and closed the lock before electricity took over. A swing bridge takes the railway line over the canal, as it continues up to **Muirtown Locks**. From here it comes within shouting distance of the **River Ness**, amidst attractive, wooded scenery.

At **Dochgarroch** both the canal and the River Ness enter little **Loch Dochfour**, which, after a spit of land shaped something like a bottle opener, gives way to the beautiful, mighty **Loch Ness**. More than 750 feet deep, it's Scotland's second deepest loch. (Only Loch Morar, at 1,000 feet, beats it.) On the southern side is **Aldourie Castle**, a classic example of Scottish Baronial architecture, with its turrets and tower. Now a hotel, it's the only habitable castle around the loch.

Loch Ness, surrounded by woods, offers many fine views of the Highlands, especially the peaks to its north. On the bank near Drumnadrochit stands the ruined **Urquhart Castle**, founded in the thirteenth century. It played an important part in the Wars of Independence in the next century, was abandoned in the middle of the seventeenth century, and was partly destroyed in 1690 to prevent it being used by Jacobites opposed to William and Mary.

Drumnadrochit itself is a small, attractive village that has become the unofficial centre of a kind of cult. It celebrates one of the most famous attractions of the Caledonian Canal. Which, alas, like the Highland weather, cannot always be guaranteed. 'Nessie', the **Loch Ness Monster**, was first spotted by St Adamnan, Abbot of Iona in the seventh century. In 1933 she posed for her first photograph. No doubt put off by the resultant publicity, this notoriously shy, Garbo-like creature, rather like the Yeti, has steadfastly maintained a low profile ever since, rarely rising to the surface. She may, indeed, have quietly made her way to a different loch altogether, in search, like so many visitors to these parts, of peace and quiet.

Further down the loch, on the south bank, is the spectacular **Falls of Foyers**, with a fall of 165 feet. On the opposite bank, further down, is the extremely pretty **River Invermoriston**.

At **Fort Augustus,** at the western end of Loch Ness, the canal now travels through a five-lock staircase surrounded by the attractive '**Canalside**'. (The siting of these locks caused Telford and his engineers much grief, after trouble from the military authorities, who had little use for a canal.)

The fort itself was built by General Wade, the Hanoverian commander in charge of military operations in the Highlands, as a response to the Jacobite Rising of 1715. It was constructed between 1721 and 1742. It was, with some prescience, as it turned out, named for the Duke of Cumberland, Prince William Augustus –the victor, or butcher, of Culloden, four years later. Before this it had been captured by Jacobites. In 1867 the fort was sold to Benedictine monks, who established a school here, **Fort Augustus Abbey**, which closed in 1993.

The canal now runs almost in parallel with the **River Oich**, again with stunning scenery. After the solitary **Kylra** and **Cullochy Locks** comes a swing bridge, and then **Loch Oich** itself. Telford had originally intended to build the canal along the southern bank, complete with towpath. Money worries led him to take it straight through the middle, having dredged the loch to create more depth.

Overlooking the loch is the

ruined **Invergarry Castle**, ancestral home of the MacDonnells, one of whom had the cheek to charge the canal commissioners £10,000 for the right to pass through his lands. Which, it should be said, are impressive, with views of woods and mountains all across the loch. A strategic castle in the days of clan warfare, it became a Jacobite stronghold during the '45 rebellion. Bonnie Prince Charlie is said to have rested here after his defeat at Culloden, whereupon the Duke of Cumberland did his best to have it blown up. Its strong walls survive. The nearby **Glengarry Castle**, built in the Scottish Baronial style in the 1860s, is now a hotel.

At the south of Invergarry is the monument of the **Well of the Seven Heads**, commemorating a satisfyingly bloodthirsty incident, part of an internecine struggle within the powerful Clan MacDonald. Two of its members were killed during a brawl, and seven of the culprits, vengefully pursued by none other than the Gaelic Poet Laureate of Scotland, a kinsman of the murdered men, were eventually killed, and decapitated. 'Bald Iain' then gathered up the heads in a plaid, washed them in the loch – where the monument now stands – and presented them to the High Chief of the Clan Donald.

After Loch Oich comes the summit level at **Laggan**. Here two steam dredgers, amongst the first ever to be used, went into action. This was engineering on a vast scale that prefigured the railway age. In *Navigable Waterways*, L. T. C. Rolt wrote: 'In sheer scale of excavation it is doubtful whether any single work of a similar kind in Britain surpasses the great cutting at Laggan between Loch Oich and Loch Lochy.' The views are equally fine.

The canal now falls to the tiny **Ceann Loch**, and then, past a spur of land, into **Loch Lochy**. As beautiful as its predecessors.

Here the River Lochy's course was diverted and dammed so as to raise the level of the loch itself to make it navigable. Telford then borrowed the river bed for the course of his own canal. In 1819 his friend and fellow poet Robert Southey (by now the Poet Laureate) came to visit the canal. Of the wholesale removal of the river, he wrote:

'Here we see the powers of nature brought to act upon a great scale, in subservience to the purposes of man: one river created, another (and that a huge mountain stream) shouldered out of its place, and art and order assuming a character of sublimity.'

Leaving Loch Lochy, the canal shadows the River Lochy, negotiating a couple of swing bridges, at **Moy** (the last one designed by Telford that remains) and **Caol**, before reaching what most visitors consider Telford's *pièce de résistance*. The great flight of eight locks, running for 500 yards, and falling 64 feet, at **Banavie**, near Fort William, with superb views down to Loch Linnhe. The longest and widest flight in Britain, they became known as **'Neptune's Staircase'**. Southey was certainly taken aback, writing that it was 'the greatest piece of such masonry in the world, and the greatest work of its kind, beyond all comparison ... the greatest work of art in Britain'.

And then, in sight of **Ben Nevis**, the highest mountain in the British Isles, come three more locks, at **Corpach**, the last a sea lock 18 feet deep. Built, with extreme engineering ingenuity, using hydraulic lime mortar, into solid rock. Here the canal joins **Loch Linnhe**, with **Fort William** to the east. The second-biggest conurbation in the Highlands (after Inverness) and a major centre for tourism, it too was named after the Duke of Cumberland, Prince William Augustus. From here the loch continues, past Oban, and the Isle of Skye, into the sea.

The biggest decision a surveyor had to make was where the summit level of the canal would be. How high it is, of course, dictates the overall number of locks the canal will need. What's more, the longer – and lower – the summit level, the more effectively rainfall can be drained off from the surrounding area. (Natural springs sometimes feed into canals, but they tend to bring in debris.) The water supply to the summit level is crucial, because every single boat travelling through it uses up two locks' worth of water, one at each end. Locks typically consume anywhere between 20,000 and 60,000 gallons of water, depending on whether they are narrow or broad. On a busy canal the amount thus lost from the summit level can run into millions of gallons a day.

Lack of water was to bedevil canals throughout their history. (And still does today.) The most effective solution was to build reservoirs, and over the course of the canal age, in its never-ending struggle to maintain enough water, more and more needed to be built. But unless a reservoir was above the summit level, and they usually weren't, water needed to be pumped up. This was expensive. So too was another water-preserving method, 'back-pumping', in which water is pumped from below a lock, or flight of locks, up to the higher level.

One canal that had to juggle with every water-saving method going was the Kennet and Avon Canal.

The Kennet and Avon Canal

Introduction by John Sergeant

I have known Bath for ages, having spent five years as a teenager at a school nearby, and I find its Georgian charm as captivating as ever. But it was only on this canal trip that I realized how Bath Stone, which gives so many of its beautiful buildings their honey colour, became so famous throughout the country. It was because of the Kennet and Avon Canal. The stone was easily cut and surprisingly hard wearing. But to reach clients clamouring to use it suppliers had to be able to ship the stone safely in bulk. And the only way to do that was to build a canal to link up with the River Thames and the wider world.

This is only one aspect of a canal which, in so many ways, deserves to be celebrated. About fifty years ago work began to bring its entire length back into use. It was Britain's biggest canal restoration project, completed in 1990. The Queen took part in a formal ceremony. In this television series, with a good deal less fanfare, I marked its successful reopening by extracting a hefty piece of Bath Stone from the last remaining quarry in the city boundary and taking it to Lancaster House in London, which has one of the finest examples of Bath stone facing in the capital. It was a wonderful canal journey, and without hesitation I put it in my top eight.

There was another aspect of its history, which made me think. I spent an enjoyable morning at Crofton Pumping Station. It's not a name which filled me with pleasurable anticipation. Even when I was told that it contained the oldest working example of a Cornish beam engine I was not as excited as I should have been. But once I had shovelled coal into the great furnace and watched with amazement as the pistons and levers sprang into action I was instantly enthralled. The steam engine was built in 1812 and sometimes carries out the task for which it was designed. In order to avoid constructing a tunnel a few miles long the engineers decided to pump water to the brow of a hill to make sure the canal could be topped up and never run dry. A tunnel was still required but would be much shorter, and you can go through it today.

The site of the pumping station has its own poignant reminder of how the Kennet and Avon met its demise. Between the great engine house and the canal you have to duck down to go through a tunnel. The bridge above still carries the main railway line from London to Exeter, built by Britain's most famous engineer, Isambard Kingdom Brunel. In 1852 the canal was bought by the Great Western Railway and with ruthless competitive logic they allowed the old waterway to deteriorate. Fortunately today we see both the canal and the railway living happily side by side.

There were all sorts of interesting moments on this journey. In summer, when the sun is shining, you see the British countryside at its best. From the canal you catch everything because your pace is so slow, and there is nothing between you, the fields and the hills. Now and then you come across a perfect country village with its church, old houses and a delightful pub. If you wish to delve into the history books to find out where battles were fought and fortunes made, there is plenty of that. But for me the Kennet and Avon Canal is an invitation to relax in one of the most beautiful parts of the country.

The Kennet and Avon, like many canals constructed at the time of the Canal Mania, found the going tough, even though its purpose – a direct route from Bristol to London – was a perfectly sensible proposition. There was, indeed, a precedent.

A canal linking the Severn and the Thames had already been built, in 1789. This was the Thames and Severn Canal, which joined up with the Stroudwater Navigation at one end, the Thames, near Lechlade, at the other. The Stroudwater ran from Framilode, on the Severn, to the wool town of Stroud. But the Thames and Severn was always in difficulties. There was a chronic lack of water, not only in the canal itself, but in the Upper Thames too. Its route, partly through oolite rock, was another major disadvantage, causing constant leakages, especially in the Sapperton Tunnel, at 3,817 yards the longest in England until 1811 (whereupon the Standedge superseded it). And the canal's western terminus was 30 miles north of the important port of Bristol.

The problems faced by the Thames and Severn – the nature of the terrain, and the inadequacies of the Thames – didn't deter the promoters of the Kennet and Avon Canal. Perhaps they should have done.

The River Avon was navigable from the Bristol Channel, at Avonmouth, to near Bath, while the River Kennet was navigable from Newbury to Reading, where it joins the Thames. A canal linking the three rivers, and thus providing a rival to the Thames and Severn, certainly worked on paper. (Although, strictly speaking, the name 'Kennet and Avon Canal' only refers to the canal that links the two rivers, the term is more usually used to denote the whole navigation, from Bath to Reading.)

Men cleaning the water in the Kennet and Avon Canal, 1942

Both the Kennet and the Avon had been made navigable earlier in the eighteenth century. As with all river navigations, objections had been made, and there were many battles with millers and landowners en route. The construction of the River Kennet Navigation, authorized in 1715, had been a particularly fraught affair, for the simple reason that the good people of Reading didn't want

Newbury trespassing on its valuable trade along the Thames. In 1720 a party of townsfolk, with the mayor at their head, had destroyed part of the works.

Meanwhile the mill-owners of the Kennet Valley, worried that the water for their mills would disappear, diverted it from the river and threw stones at the canal-builders. Compensation for damage caused by the navigation's construction was, eventually, paid. It was fully open by 1724.

But the Reading bargemen were still less than enthusiastic about their new neighbour. In 1725, a blood-curdling letter was received by a Maidenhead barge-master who had had the temerity to use the new waterway.

Wee Bargemen of Redding thought to Aquaint you before 'tis too Late, Dam You, if y. work a bote any more to Newbery wee will Kill You if you ever come again this way, wee was very near shooting you last time, wee went with pistols and was not too Minnets too Late.

After much opposition work had begun on the Avon Navigation in 1724. It was successful, and coal was imported, via the Severn, from the Shropshire coalfields, much to the fury of Somerset coal-miners, who found they were being undercut. (The Somerset Coal Canal was later built to connect to the Kennet and Avon Canal, doing so in 1801.)

The 'Western Canal', as the Kennet and Avon was first known, was proposed in 1788. The driving force behind it was Charles Dundas, MP for the County of Berkshire, and associates of his in Hungerford and Marlborough. Robert Whitworth, who had learned his trade as one of

Brindley's 'School of Engineers', was asked to carry out an initial assessment.

Whitworth had done the original survey for the Thames and Severn Canal. Not wishing to have his fingers burned again, he concluded there was insufficient water available for the Western Canal to be viable. His services were dispensed with.

In 1790 the twenty-nine-year-old John Rennie was appointed to make a further survey.

John Rennie went on to become one of the great names of civil engineering. But the Kennet and Avon was only the second canal he was involved in, the first being the Lancaster Canal, on which he was engaged at the same time. (The Lancaster Canal is noted for Rennie's Lune Aqueduct, one of the most celebrated canal constructions.) He went on to work on the Crinan and Rochdale Canals.

Like Watt and Telford, Rennie was a Scot, born in 1761, in East Lothian. The son of a farmer, he spent a good deal of his spare time at the mill-wright business of Andrew Meikle, the inventor of the threshing machine. Unlike the other great names in civil engineering at the time, he attended university, at Edinburgh. But he spent his vacations working at the mill.

In 1784, like William Murdock before him, Rennie walked down to Birmingham and presented himself to James Watt at the Soho Manufactory. Impressed, Watt offered him a job. Rennie was sent to supervise the building of a massive steam engine at Albion Flour Mills at Blackfriars. This was London's first great factory, thought to have been the inspiration for William Blake's 'dark satanic mills' and one of the first attempts to use steam to drive flour mills. (When they were built, Erasmus Darwin called them 'the most powerful machines in the world'. They didn't please everyone. In 1791, five years after they opened, the Albion mills were burned down by disgruntled rival mill-owners, disgruntled mill-workers, or both.)

By this stage Rennie had picked up a good grounding in construction, steam and iron. The clever use of this metal would become a signature mark of Rennie's bridges, as it was in those of Telford. (Rennie later built Waterloo and Southwark Bridges, as well as designing London Bridge, now in Arizona.)

His Achilles heel would prove to be his handling of water supply. Rennie conducted his own survey for the proposed Kennet and Avon Canal – and came up with the answer the committee wanted to hear. Water, he assured them, would

John Rennie
(1761–1821), the
Scottish engineer

not be a problem. In 1790 the promoters began the process of getting the canal set up and authorized.

But by now, Canal Mania was in full throttle. Speculators were falling over themselves to subscribe to a new canal – any canal.

In 1791 a group of chancers met in the White Lion Inn in Bristol and tried to raise money for what was in effect going to be a takeover of the Western Canal. Wanting to keep the subscriptions, and thus future shareholdings, to themselves, and fearing they would be swamped by rival speculators, the promoters took to subterfuge. Counting on the poor road transport available in December, they took out an advertisement claiming that a meeting to discuss a canal from Bristol to an as yet unspecified point in Wiltshire would be held in Devizes – only two days after the notice appeared.

The plan backfired, badly. Intrepid investors realized that this must have something to do with the proposed Western Canal and large numbers of them feverishly battled their way to Devizes, only to find there was no such subscription meeting taking place. The whole thing had been a charade. Fearing a riot, the town clerk insisted a meeting went ahead. A Bristol promoter who was present, thinking quickly, now proposed a completely new canal, from Bristol to Salisbury. Remarkably, the idea was warmly accepted by the gathering, who happily dispersed. Such a canal was never built.

The 'White Lion Junta' (as they were called) now made peace with Dundas's Hungerford faction. The money was raised, and plans for the canal went ahead.

The proprietors of the Avon Navigation, once they got

wind of the scheme, were keen to take a slice of the action and in 1793 offered the waterway for sale. This was declined, but when the Kennet and Avon Committee changed its mind, the offer had been withdrawn. In secret, the canal's promoters began to buy up shares in the Avon Navigation, purchasing them through intermediaries. They had secured a controlling interest by 1796 and were eventually to own all but one of the shares. Thus they could exert pressure on their neighbour. (One way they did so was to threaten to build a brand new canal that would take over much of the Avon Navigation's traffic unless the proprietors woke their ideas up and built a badly needed towpath.)

At first the committee members weren't sure whether they wanted to build a broad or narrow canal. Wisely, they settled on the former.

An Authorizing Act was secured in 1794, and work could begin in earnest. But by now Rennie had already realized that he had been over-optimistic in his initial findings. There would, he now saw, be considerable difficulties with water. His solution was to move part of the route further south. But this would mean building a 2½-mile tunnel. William Jessop now became involved and suggested yet another route, which would cut the period of construction down by two whole years and save £47,000. (The route was subsequently altered on several more occasions. Not least because business inter- ests in Devizes had managed to pressure the canal committee into bringing the canal near to their town.)

But the savings on the tunnel would be at the cost of a relatively short summit level. This would mean that not only would a reservoir be needed, but so too would steam pumps, to draw the water up to the level of the canal. (Two pumping stations were indeed built, at Crofton and Claverton.)

There were huge challenges. The summit pound would be 452 feet above sea level. Though this was child's play by the standards of the Huddersfield Narrow Canal, whose Pennines summit level at Standedge stood at 645 feet, it was still a considerable undertaking. The canal was to be 57 miles long, with 79 locks, and accommodate craft up to 70 feet 10 inches long and 13 feet 8 inches wide.

Once construction began, further difficulties emerged. A full geological survey hadn't been carried out, and it was now discovered that large parts of the canal's middle section would have to be dug through hard rock, the presence of which had remained undetected, under what turned out to be only a shallow bed of clay. A knock-on effect of this was that there was not enough clay available for the bricks needed for construction, and so, at great expense, they had to be bought in.

The canal, like its northerly neighbour, the Thames and Severn, was dug through oolite and chalk, which leaked water and provided a constant headache, requiring endless remedial work, all of which came at a hefty cost.

In a rush to get things finished, both workmanship and materials were skimped on, with the result that several constructions on the canal had to be rebuilt. The Avoncliff and Dundas Aqueducts, two of the great glories of the canal, were flawed from the outset. Bath Stone, in which that butterscotch-coloured Georgian city is largely built, needs to be carefully weathered before being used for construction. If not, cracks appear. As was the way with other parts of the Kennet and Avon, short cuts were taken in order to save money. The Bath Stone used for the aqueducts wasn't properly weathered. Cracks did appear.

Ironically, given that the aqueducts are amongst his most admired creations, Rennie had constantly urged the company to let him build them in brick rather than stone, foreseeing the problems. But he was overruled – on grounds of cost.

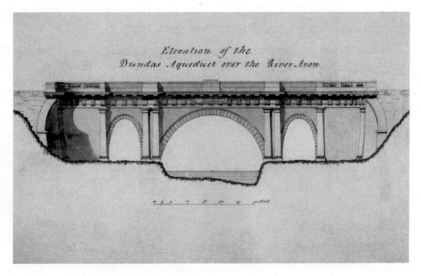

Rennie's design for the Dundas Aqueduct, *c.* 1799

All this time, the Napoleonic Wars were pushing up prices, and that, along with the unforeseen delays, meant that the canal was constantly running out of money. Work was often stopped until more cash could be raised. Not until 1810 was the Kennet and Avon finally opened.

Another difficulty was posed by the Thames itself. The river had few towpaths. Where barges couldn't rely on sails, groups of 'bow-haulers' manhandled them along the river. There were several of the dangerous and inefficient flash-locks still in operation. The pound-locks that were in place left a good deal to be desired too.

Ever since 1197 the Thames up to Staines had been in the control of the Mayor and Corporation of London. Conservancy on the rest of the river was the responsibility of local authorities. As with roads, their stewardship, especially of the upper reaches, was often dilatory. In the second half of the eighteenth century, though, steps were taken to improve matters, the river upwards of Staines eventually being placed in the control of the Thames Navigation Commissioners in 1771.

By this stage James Brindley himself had been called in to advise the Corporation of London on how to improve the river. His solution, naturally, was a canal – the 'London Canal' – which would run from Sonning, near Reading, past Monkey Island, near Maidenhead, to join the Thames at Isleworth. The canal would have taken vessels loaded with up to 200 tons of freight. Rejected by parliament, it was never built.

As always with river navigations, objections were raised to any attempts to make improvements on the Thames. Millers, as usual, were strenuous in their opposition. People who owned land and houses on the river also lobbied against attempts to build towpaths and stimulate water trade.

The great canal enthusiast John Phillips, in his *General History of Inland Navigation* of 1805, sums up the waterway lobby's feelings about the nimbyism of these wealthy parvenus:

Those fine gentlemen would not suffer their villas to be disturbed by noisy boatmen, or their lawns to be cut through for the accommodation of trade and Commerce, though it was from that only that most of those villas and lawns had existence.

After this, piecemeal improvements were made. New locks were built, including Boulter's Lock, near Maidenhead. But not enough was done to really make a difference. By the 1790s a stand-off had developed because the Thames Commissioners didn't want to shell out for improvements until they saw signs of increased trade, while the canal companies weren't about to increase trade until improvements were made.

More London canals were proposed, but the bigwigs of the Corporation of London were far too canny a set of operators to allow such upstarts to put paid to their river traffic. Schemes for new canals were quietly strangled at birth, after the usual flurry of lobbying, arm-twisting, ear-bending and palm-greasing behind the scenes.

Only in the early years of the nineteenth century were serious attempts made to take matters in hand, and bigger locks were finally put in place, improving the speed of long-distance trade.

Closer to home, in 1812 the Kennet and Avon Company got authorization to buy the Kennet Navigation. In the same year it finally persuaded the Avon Navigation to build a towpath.

The Kennet and Avon was never a real money-spinner.

Its dividends were relatively low because of the costs of construction and maintenance. Trade between Bristol and London had declined by the time it was built. As there was little in the way of large-scale industry most of the traffic, more so than elsewhere, was local. Arguably, its perennial lack of water supply would have precluded a greater volume passing over the summit in any case.

Nevertheless, there was a valuable source of trade in Bath Stone, which was transported to London. Large shipments of coal and stone came up the Somerset Coal Canal which joined it east of Bath around 1800. Ten years later, the navigation hooked up with the Wilts and Berks Canal near Melksham, from where it made its way to the Thames at Abingdon. These improved transport links lowered the costs of goods and made more of them available, thus increasing the prosperity of the towns served. Toll receipts nearly doubled between 1812 and 1823.

(Ambitious plans to build canals linking the Kennet and Avon to the Basingstoke Canal, and from Reading to the Grand Junction Canal near Uxbridge, did not come off.)

The Kennet and Avon experienced a bizarre coda during the Second World War – but not as a waterway. It became part of the fight against Hitler. France fell to the Germans in June 1940 and England seemed to be next on their itinerary. Plans were rapidly put in place, not so much to halt a possible German invasion, as to harass and delay it. As Churchill put it, in his famous speech that same month:

we shall defend our island, whatever the cost may be. We shall fight on the beaches, we shall fight on the landing grounds, we

shall fight in the fields and in the streets, we shall fight in the hills;
we shall never surrender.

What Churchill didn't add was that we would fight on the canals too.

The Kennet and Avon Canal was called up. No longer really a going concern as a waterway, and thus unsuitable for military transport, it would now become one of over fifty lines of fortifications that were to form a military corset around Britain. Rivers and railways were also drafted in.

The most important, and the longest line, was the 'General Headquarters', a 'stop line' or anti-tank line, which was to act as a southern barrier arcing from Bristol to London. (The GHQ Line was of course, in part, the province of 'Dad's Army', the Home Guard, which had been formed in May 1940.) After the fortifications on the south coast, this was the last line of defence standing between the invaders and their prize. The Kennet and Avon became part of 'Stop Line Blue'. Troops were sent in to lay barbed wire and build pillboxes, anti-tank obstacles and gun emplacements. If the Germans did invade, and could be ensnared among these defences, the army – or so it was hoped – could counter-attack. Several pillboxes are still to be seen along the canal.

Having done its bit, the next chapter for the Kennet and Avon was thoroughly dismal. After the war, its decline was vertiginously rapid, as the British Transport Commission tried its best to put it out of its misery and hasten its final demise. A canal section could be closed if there was no traffic on it for a year, and the BTC set about trying to make sure this was the case. Not only by allowing the waterway to become dilapidated and choked up with weeds, but by sneaky tricks such as colluding with Reading Corporation in making permanent repairs to Bridge Street Bridge, which dramatically reduced the headroom, making navigation impossible for all but the smallest boats. By 1954 only a few sections of the canal were passable. The next year, the BTC applied to the government, asking permission to wash its hands of the canal altogether.

But then the tide turned. Strong protests were made against the BTC's plans. (At a subsequent court case brought by the few remaining carriers, a hostile witness had asked, 'Is it right to spend hundreds of thousands of pounds on a canal merely because one man wants to sail a couple of boats on it?' He couldn't have more summed up the BTC's attitude to canals more succinctly.)

Parliament threw out the BTC's Bill. But the dander of all people who valued the canals was up. The subsequent campaign to restore the Kennet and Avon was one of the turning points in the post-war history of the movement to revitalize our canals.

The last remaining commercial carriers had fought tooth and nail to keep it open, and now a host of pleasure craft owners, and hirers, joined in, literally forcing their way through the closed sections and putting pressure on the BTC and local authorities to abandon their plans to let the waterway die.

Restoration work began, and by 1966 the Kennet and Avon Trust had restored and reopened Sulhamstead Lock, on the Kennet Navigation. In 1990 HM the Queen formally reopened the whole waterway, forty years after it was last fully navigable.

The Kennet and Avon Canal
– the journey itself

Moorings at Avoncliff

The Kennet and Avon may not have been a rip-roaring success as a commercial concern, and it still suffers from its perennial shortage of water, but thanks to the dedicated band of men and women who refused to let it dry up altogether, it has survived. Today it is celebrated as one of the most beautiful of our inland waterways, with its long stretches of peaceful countryside, market towns, elegant bridges and aqueducts and, of course, the World Heritage city of Georgian Bath. One of the reasons it failed to attract large volumes of traffic – the fact that it passed through relatively little industry – is today, from a leisure point of view, one of its great strengths.

Should you choose to do the whole journey, the two rivers as well as the canal between them, you could start at Avonmouth and then make your way eastwards, through the spectacular **Avon Gorge**, under the elegant **Clifton Suspension Bridge** (designed by Isambard Kingdom Brunel) and into the city of **Bristol**, with its famous **Floating Harbour**. Engineered by William Jessop, opened in 1809, the harbour was saved from destruction by the lobbying of the Inland Waterways Association and local campaigners in the 1970s and has since been restored. The attractive waterside is now home to restaurants, bars and galleries.

Moored nearby are Brunel's famous SS *Great Britain*, the world's first ever iron-hulled, steam propeller-driven ocean liner, and a replica of the Matthew, in which the Italian adventurer Giovanni Caboto – known to history as John Cabot – sailed from Bristol to a still unidentified landfall in North America in 1497. On his way, he hoped, to Asia by a quicker route than that chosen by Columbus five years earlier. He was to be disappointed.

The tidal River Avon gives way to the **Avon Navigation** at **Hanham Lock**. The canal supervisor's house is still standing, a Grade II listed building. In Hanham itself (once a village, now part of Bristol's urban sprawl) is the **Blue Bowl**, said to be Britain's oldest tavern, thought to date from Roman times. From here the Avon passes through pleasant woodland, oblivious to the Bristol suburbs further out. On the north bank is **Cleeve Wood**, a designated Site of Special Scientific Interest, home to the rare and fabled Bath Asparagus.

The first incarnation of **Keynsham Lock** was built in 1727. Nearby is the church of **St John the Baptist**. There is a musical connection. When George Frederick Handel first heard the organ here, he liked its tone so much that he bought it, the price being a set of eight bells, still there.

Further up the river the canal is shadowed, and crossed, by one of its old rivals, the Bristol and Bath Railway, its line now mostly a walking and cycling path. The **Avon Valley Steam Railway**, staffed by volunteers, operates from a station at 'Avon Riverside'.

Saltley Lock (also built in 1727, but destroyed by disgruntled Somerset coal miners in 1738, unhappy that cheaper coal was being brought by waterway from Shropshire and putting them out of business) has an impressive weir, as well as a marina and a thatched riverside pub, the **Jolly Sailor**. From here the river rises to meet Bath via a series of seven locks, the **Widcombe Flight**. **Thimble Mill** was once a pumping station, built to aid the navigation's water supply.

After **Newbridge** the suburbs of Bath make their presence known. The old city centre of **Bath**, with its creamy golden stone, is one of the best-preserved Georgian cities in Britain (and Ireland), with its famous Royal Crescent, Circus, Gay Street and Queen Square (the last three shaped so as to resemble a masonic key). Bath was Pleasure Central from the moment Queen Anne visited the spa in 1702, declining only at the end of the century as the new craze for dipping in the briny at Brighton took over. Still surviving from its Georgian heyday are the famous Pump Room (with its sulphurous healing waters, built above the Roman baths of Aquae Sulis) and the beautiful Assembly Rooms (home to the eighteenth-century marriage market). After Halfpenny Bridge, at **Bath Bottom Lock**, the canal splits off from the river, never to rejoin it.

The River Avon itself passes under the Italianate Great Pulteney Street Bridge, with its twin arcades of tiny toy-like shops, unique in Britain, designed by Robert Adam. It was named after Sir William Pulteney, Telford's protector, who built a new town on the other side of the river.

The canal, meanwhile, has taken a separate path to the east of the river and made its way to Bath's most important 'pleasure garden', at the far end of Great Pulteney Street, **Sydney Gardens**. (Jane Austen once lived nearby.) Pleasure gardens were all the rage in Georgian Britain, open to all but the very poorest classes, and a centre of music and entertainment and social life. Every town worth the name had one.

And while Bath (though it didn't know it) was on the wane as *the* resort of fashion by 1795, when the Gardens were opened, the canal company had to pay through the nose for the privilege of cutting right through the centre of them in 1810. Two thousand pounds changed hands. John Rennie rose to the occasion and made the canal into an attractive 'water feature', with its elegant iron bridges and two tunnels (both built for aesthetic rather than practical reasons).

Thirty years later, Brunel's Great Western Railway followed suit (and altered the course of the canal when it did so). Sydney Gardens, still popular with Bathonians today, is the last remaining pleasure garden in the country in anything like its original condition. On the site of what was once the grand entrance is the **Holburne Museum**, an art gallery.

As it leaves the Gardens the canal passes underneath **Cleveland House**, the former HQ of the Kennet and Avon Canal Com-

pany. A trap door was built into the tunnel it stands on so that paperwork could be exchanged with the boaters below.

Bathampton Meadow and the area surrounding this stretch of canal is rich with flora and fauna, especially birdlife. Here the famous Bath Stone was quarried and shipped to the city along the Avon, where it was taken to the centre by tramroad. After the canal leaves Bathampton, it's back in open, hilly countryside.

Still housed in its attractive Georgian building, the **Claverton Pumping Station** was built by John Rennie to draw water from the River Avon for the canal, 48 feet above. It began doing so in 1813. The station was restored by volunteers, work beginning in the 1960s, and was once again operational by 1978.

Now entering Wiltshire, the canal continues on its separate course, sometimes within hailing distance of the Avon, sometimes far away. It crosses and recrosses the river in the two splendidly designed, if materially flawed neo-classical aqueducts that Rennie built with Bath Stone. **Dundas Aqueduct** is faced with sets of twin Doric pilasters.

Avoncliff Aqueduct has a Corinthian entablature. Past Avoncliff is **Barton Tithe Barn**, at 168 feet one of the longest in Britain.

The Dundas Aqueduct was the junction where the Somerset Coal Canal joined the Kennet and Avon. It's now disused, apart from a stretch used for mooring at the wonderfully named **Brass-knocker Basin**.

The first sod of the Kennet and Avon canal was cut at the former wool town of **Bradford-on-Avon**, outside the **Canal Tavern**, in 1794. This attractive town was largely built in Bath Stone. Parts of the historic wharves still stand. The town bridge, with its chapel (people in medieval times thought it a good insurance policy to pray before crossing a bridge), are Grade I listed buildings. (The chapel later underwent a significant 'change of use' – as a gaol.) The Saxon church of **St Laurence** was only unmasked as such in 1857.

After aqueducts carrying the canal over the **Rivers Biss** and **Semington** comes **Buckley's Lock**. Here a lock-keeper's cottage is the only relic of the disused Wilts and Berks Canal,

which once went on from here to join the Thames at Abingdon. From Semington the Avon parts company with the canal once and for all. At **Seend** the water-way passes by the garden of the cunningly named **Barge Inn**.

At **Foxhangers** the canal now begins its steep climb up to Devizes. At **Caen Hill** is a famous flight of sixteen locks, described by L. T. C. Rolt as 'the most spectacular lock flight in England because the locks are broad and laid out in a straight line so that they can be seen in perspective'. To help preserve precious water the locks were built with enormous side-ponds. It's one of three flights over a 2½-mile stretch, totalling twenty-nine locks in all, rising nearly 240 feet.

Several fine historic buildings survive at the once busy hub of **Devizes Wharf**, including what are now homes to the **Wharf Theatre** and the **Kennet and Avon Canal Museum**. For beer lovers, there's the visitor centre of the **Wadsworth Brewery**, which supplied ale to the numer-ous pubs along the canal. Many, sadly, now vanished, as is the case on all the canals. A wartime pillbox still guards the wharf. Devizes was an important market town – as evidenced by its Georgian Town Hall and Victorian Corn Exchange.

East of Devizes the canal now travels through the beautiful **Vale of Pewsey**. Along with the Marlborough Downs, Savernake Forest, White Horse Hills and Berkshire Downs, all of which the canal will encounter, the Vale of Pewsey is part of what is now designated the **North Wessex Downs**, an official Area of Outstanding Natural Beauty. In this section are many of the Kennet and Avon's distinctive swing bridges, built as 'accommodation bridges' so farmers could access their fields.

As you pass along to the Vale of Pewsey, south of the North Wessex Downs, you will see the prancing prehistoric **White Horse of Uffington**, 374 feet long, its sinewy contours picked out of chalk.

The Kennet and Avon now passes along flat countryside until it reaches the pretty canal village of **Honeystreet**. Here (as at Pewsey and Wootton Rivers) is a mixture of brick, thatched and half-timbered buildings. Another **Barge Inn**, threatened with closure, was bought by the locals. Its name recalls that this was once a busy boat-building centre. Further along the canal, near **Church Farm Lane Bridge**, are the remains of some Second World War anti-tank cylinders. From here towards **Pewsey** is a series of simple but pleasing stone bridges.

Before Wilcot there is an exception to the rule: a self-consciously elegant balustraded bridge, known as **Lady's**, or **Ladies Bridge**. The landowner here, Lady Susannah Wroughton, as well as snaffling £500 for permission for the canal to pass through her property, also insisted on this decorated bridge being built. She further demanded that the canal widen and turn itself into an ornamental lake, which still survives as **Wilcot Wide Water**.

Pewsey Wharf is a small, attractive canal centre with a waterfront pub and café. In the village is a statue built in 1913 to commemorate a former landowner here. He's Alfred the Great, the famous cake-burning, Dane-defying King of Wessex, later of all the Anglo-Saxons.

Near the village is a fifty-year-old land-carving of a white horse. We are now in crop circle territory.

From **Curret Crown Bridge**, with only rare detours, the old GWR line runs fast alongside the canal it came to own, right up to the outskirts of Newbury.

At **Wootton Rivers**, near Pewsey, relics of Second World War pillboxes and a gun emplacement can be seen. Several former canal properties still stand, converted for residential use. The village buildings are mostly thatched.

The Vale of Pewsey now gives way to the ancient **Savernake Forest**, 'improved' by Capability Brown, with its oak and beech, as well as softwood trees planted for sale as timber. This is the site of the Kennet and Avon Canal's summit level.

The **Bruce Tunnel** at the Savernake summit – the only one on the canal – is 502 yards long and 17 feet 4 inches wide – and tall, at 13 feet 2 inches. When it was built, chains were fixed to the walls so that boats could be hauled through. Somewhat ironically, it is called the Bruce Tunnel after the family name of the Earl of Ailesbury. He owned the land on which it was built and insisted a tunnel was built to hide the canal out of sight, instead of the deep cutting the engineers had actually planned.

After it emerges from the tunnel, the canal passes through very wonderful scenery, with many brick bridges that fit in well with the character of the countryside, despite the railway line's threatening presence.

Wilton Water was dammed to become the main reservoir for the canal. The sluices and outfall are listed buildings. The water was pumped up into the canal by the two Boulton and Watt steam engines (the last such still in working order) at **Crofton Pumping Station**. It was restored in 1970 and opened by Sir John Betjeman, crown prince of the movement to preserve and restore old buildings. The canal now descends towards Hungerford, through a flight of locks at Crofton.

John Rennie was the first engineer to build skew, or diagonal, canal bridges. Examples can be seen at **Beech Grove** and **Mill Bridge**. The canal runs past pretty lock cottages towards **Great Bedwyn** and **Little Bedwyn**. Here it passes through flat farmland, both arable and pasture, towards **Froxfield**, the border with Berkshire. In this county the Kennet and Avon will stay until it reaches Reading, the county town.

An aqueduct takes the canal over the River Dun, past the wildlife haven of **Hungerford Marsh**, with its two swing bridges, and into the old market town of **Hungerford**, where the idea for the Kennet and Avon Canal first emerged. The scenery from here down to **Kintbury**, past water meadows, is particularly fine. We pass the Norman church at **Avington**, with the River Kennet nearby.

In the village of Kintbury the canal is overlooked by the **Dundas Arms**, named after the first chairman of the Kennet and Avon company, Sir Charles Dundas, MP for the County of Berkshire. He lived at nearby Barton. From here to Newbury, past **Shepherds Bridge**, through **Dreweat's Copse** and **Hamstead Locks**, with wooded hills, water meadows and **Hamstead Park** to the south, is another celebrated stretch of scenery.

Downstream of **Newbury**

Lock is a stone bridge where there was no towpath, and the horse had to be uncoupled, the line being floated under the bridge by means of a buoy. At **Newbury Lock Cottage** an antique notice still warns anyone tempted to ignore this procedure that 'The Captain of every vessel allowing Horses to Haul across the street will be Fined.'

At the former cloth town of **Newbury** the townspeople seem to have been rather more welcoming to the navigation than were those of Reading. It runs right through the centre of the old market town. Although the old canal basin was idiotically filled in and converted into a car park, several old wharf buildings remain from what was once a busy trading hub. The **Stone Building** is the headquarters of the Kennet and Avon Trust. An antique crane stands sentinel.

The canal proper comes to an end at Newbury, and the **River Kennet Navigation** takes over, past **Newbury Racecourse** towards Reading. In the 1720s 11 miles of new cuts were made, along with 20 turf-sided locks. The extensive restorations to this section of the waterway are eloquent testimony to the efforts and determination of the Kennet and Avon Trust and local authorities (with the help of job-creation schemes). **Swing bridges** are a particular feature of this section. Rennie designed them to be mounted on ball-bearings, an early use of the technology. Many locks and swing bridges have been restored in recent years.

Towards Thatcham, surrounded by peaceful water meadows, **Monkey Marsh Lock** is to be found – one of only two of the remaining turf-sided locks.

At **Woolhampton** is a preserved pillbox, dating from the canal's distinguished war service. Up on the hill above the town is the Roman Catholic **Douai Abbey** and School. At **Aldermaston Wharf** there is a hydraulic lift bridge carrying the busy road to Basingstoke over the canal. The lock here has distinctive scalloped sides. At **Padworth Lock** a former canalman's cottage is now a Kennet and Avon Trust **visitor centre**.

Downstream, a section with attractive woods and arable land leads to **Sulhamstead Lock**. Its restoration in 1966 was one of

the triumphs of the inland waterway restoration movement. At Sulhamstead is one of the many distinctive Rennie swing bridges.

Sheffield Lock also has scalloped sides, while **Garston Lock** is the other surviving example of a working turf-sided lock. From here to **Fobney Lock** (apart from when the M4 crosses over the navigation) the countryside offers fine scenery. But then the urban landscape of Reading takes over, as the Kennet makes its way to join the Thames below Caversham.

Reading was famous in turn for its cloth, seeds, railways and biscuits. On the waterside only one building remains of what was once the largest biscuit factory in the world – **Huntley and Palmers**. There are also a museum and other new developments at the old wharves, making this an attractive and smart waterside area. The **Oracle** shopping and leisure mall is built on the sites of a seventeenth-century workhouse of the same name and the former Simonds Brewery (which was taken over by Courage). The town's other attractions include its minster church of **St Mary the**

Virgin, its ruined **Abbey** (built by Henry I) and the **Maiwand Lion** in Forbury Park, a bristling 31 feet of iron defiance, built to commemorate (British) lives lost in the Second Afghan War.

The Kennet Navigation joins the **Thames** after passing through **Blake's Lock**, by the site of the old Reading gasworks. From Reading the river winds its way down to London, passing through the urban corridor of Twyford, Maidenhead and Slough, but managing, miraculously, to keep out of sight of most modern industrial developments and remain rural in character for long stretches around Henley, Marlow, Cookham (especially Cliveden Reach), Bray, Dorney, Boveney, Datchet and Runnymede, before passing through the Surrey suburbs up to Brentford (once in Middlesex), and then London Town.

The Evolving Canal

The Railway Threat

The Stockton and Darlington Railway, opened in 1825, was something of a half measure. On some sections the carriages were pulled by stationary steam engines, while passenger traffic was horse-drawn. The real pioneer of the railway age, in which the steam locomotive was king, was the Liverpool and Manchester Railway Company. Founded in 1823, it secured its Authorizing Act of Parliament three years later. George Stephenson was its engineer.

The line was proposed because of the exorbitant tolls being charged by the Bridgewater Canal and its old foe, the Mersey and Irwell Navigation. Fearing that this virtual monopoly of transport between the two cities was hobbling their industrial development, wealthy merchants and manufacturers got together and decided to build a railway. The Duke of Bridgewater's nightmare, the 'damned tramroads', had arrived.

When Stephenson's surveyors came to look at the lie of the land, the superintendent of the Bridgewater Canal placed armed men around the estate to try to keep them out. This was a bit rich, as canal surveyors had, in the past, surreptitiously surveyed land by night. (And, on one occasion at least, the property of a vicar on a Sunday morning, while he was officiating at church.)

The activities of Stephenson and his surveyors – looking not only into the Liverpool to Manchester line but also into potential links between London, Birmingham, York and south Wales – alerted the other canal commandants to the threat posed by the oncoming juggernaut. After decades of doing very little to improve their canals, they were at last stung into action.

At the centre of these belated attempts to upgrade the canal system were two men, Thomas Telford and James Brindley. Telford began a whole series of works to improve Brindley's canals. In 1824 he was asked to survey the Birmingham Main Line, still the busy hub of the whole network. In a famous memorandum, he reported what he'd found:

> *Where it enters Birmingham, it has become little more than a crooked ditch with scarcely the appearance of a haling-path [hauling-path or towpath], the horses frequently sliding and staggering in the water, the haling-lines sweeping the ground into the canal, and the entanglement at the meeting of boats being incessant. Whilst at the locks at each end of the short summit at Smethwick, crowds of boatmen are always quarrelling or offering premiums for a preference of passage, and the mine-owners, injured by the delay, are loud in their just complaints.*

Not only had the canal fallen into disrepair. Brindley's and Simcock's luxuriant S-bends had infuriated canal carriers and boatmen for more than five decades. James Watt himself was one of many people who'd proposed making cuts in them.

Telford's solution was drastic and dramatic. He made a cutting 70 feet deep at its maximum and 40 feet wide and built a new, straight main line, with towpaths on each side. It was finished in 1829. The original Brindley/Simcock lines were left as feeders.

Further cuts across other loops in the Black Country were made. By the time they were finished, in 1838, Brindley's original route of 22 miles from Birmingham to Tipton had been cut down to 15 miles – lock-free.

Even the Oxford Canal was persuaded to stir itself out of its customary lethargy. Brindley's languorous journey though the fields of Oxfordshire had long been a cause of frustration. Finally, between 1829 and 1834, the notorious stretch of the route between Hawkesbury and Wolfamcote was shortened by 14 miles, thanks to the Telford-like technique of bold cuttings and embankments (although the great man was not, on this occasion, involved).

The Trent and Mersey Canal also began to make improvements. The main bugbear was Brindley's Harecastle Tunnel. Not only was it a notorious bottleneck, it had not been properly maintained. When John Rennie was asked to report on it in 1820 he found that the brickwork inside was so badly affected by the wear and tear of passing boats that bricks could be pulled out by hand. In places the tunnel roof was only six feet from the waterline. The company had done nothing to rectify matters.

Four years later Telford was hired to build a new tunnel right next to the old one. Such were the great advances that had been made in engineering in the intervening fifty years that the new tunnel took just three years to build, as compared with eleven for the original. And the new tunnel had a towing path.

Both were used, boats coming north taking Telford's tunnel, those travelling south using Brindley's.

The Macclesfield Canal, laid out by Telford, and built between 1826 and 1831, could have been a valuable addition to the network had it been built ten years earlier. It runs from near Harecastle north to the Peak Forest Canal, from which links can be made to the Bridgewater, Rochdale, and Huddersfield Narrow Canals.

Telford's last major canal project was one he had suggested himself, the Birmingham and Liverpool Junction Canal. Backed by the Birmingham Canal Navigations, it received authorization in 1825.

There is no nonsense about the Birmingham and Liverpool Junction. Typically for Telford it drives on in as straight a line as possible, achieved, as usual, with the help of mighty earthworks and cuttings along its route. As L. T. C. Rolt remarks:

> *No-one can travel through this canal without being impressed by the contrast between it and the earlier contour canals engineered by Brindley and his school. It emphasizes the progress made in civil engineering during the canal era, illustrating that 'cut and fill' technique which would soon be used by the railway engineers and, in our own day, by the builders of motorways. It is significant that Telford's last canal is still the most direct route between the Black Country and the Mersey.*

(In some cases, though, Telford overplayed his hand and the embankments were subject to slippage.)

The Birmingham and Liverpool Junction joins the Staffs and Worcs at Autherley, near its junction with the Birmingham Main Line. From Autherley it runs north to Middlewich. And here, at last, the Ellesmere and Chester Canal Company (now with powerful allies) finally managed to connect to the Trent and Mersey Canal, after nearly sixty years of being forbidden to do so.

(Lest it be thought that the canal companies, faced with the threat of the railways, were at last starting to show some wider instinct for survival and beginning to cooperate with each other, it should be noted that the Trent and Mersey Company insisted on building its own canal between the two main lines. The Wardle Canal, at 154 feet Britain's shortest, was by no means its cheapest. The Trent and Mersey demanded stiff tolls from the Ellesmere and Chester Company for the privilege of using it.)

There was one very notable success story from this period of frantic, railway-inspired improvements to the system. In 1827 a canal was completed that was to be one of the few that remained commercially viable come the final death throes of the canal system that began in the 1960s. This was the Gloucester and Berkeley Ship Canal – now called the Gloucester and Sharpness Canal after its later terminus on the Bristol Channel.

It had first been authorized in 1793, at the height of the Canal Mania, but, as was the case with many projects in that benighted period, it proceeded only in fits and starts. Only in 1817 did it really get going. Significantly, this was because it received help from central government, which provided money to alleviate unemployment after the Napoleonic Wars. Telford was chosen to survey the final route.

At the time of its construction, the Gloucester and Sharpness was the broadest and deepest canal in the entire world. At 86 feet wide and 18 feet deep – many canals were nearer 4 – it could accommodate not just coastal, but ocean-going ships. It was made even larger at various points in the future, as ships grew bigger. Its sheer size was to give this gigantic waterway a significant advantage in the future.

Here are a whole series of *what might have beens*. From now on, until as late as the 1940s, eyes turned longingly to the possibility of finishing what had been started in the 1820s: ship canals, new, direct routes such as the Birmingham and Liverpool Junction and Macclesfield Canals, and the widening, straightening, deepening and upgrading of the original network. Had this work been undertaken when money was flooding into the canal companies' coffers, their future history may have been different, and the railways might have been prevented from riding roughshod over the whole system.

However, the companies were sloughing off the long-accumulated threads of their complacency and indolence to no avail. Much-needed improvements had been made. But they were too little, and too late.

George Stephenson's locomotive, the *Rocket*, 1829

In 1829 the Liverpool and Manchester Railway Company held the famous Rainhill Trials, offering a prize of £500 for the best steam locomotive. The winner was George Stephenson's own engine, the famous *Rocket*, which achieved the remarkable top speed of 30 mph.

The line opened on 15 September 1830. The prime minster, the Duke of Wellington, was persuaded to travel in one of the eight inaugural trains. But the day was hardly an auspicious success. One of the trains derailed, and the one following collided with it. Worse was to come. The MP for Liverpool, William Huskisson, formerly a go-ahead president of the Board of Trade, went over to the duke's carriage, hoping to shake his hand. Suddenly realizing that another train, the *Rocket* itself, was coming up fast on the adjacent track, he panicked and tried to clamber into the carriage. But the door swung open and he fell on to the track, and his leg was run over. He died later that night.

These delays meant that the large crowd spilling on to the rails at Manchester waiting for the trains to arrive got restless and pelted the duke's carriage with rotten fruit when it finally reached the city. The eventful trip had seen the first widely reported railway fatality – and the first of many delays.

Despite these teething problems, the line became a success. Surprisingly for its owners, though, its first main cargo was passengers.

But if the canal companies breathed a sigh of relief, they were wrong to exhale. Freight was soon to follow the passengers' lead. (And passengers were, in any case, more profitable a cargo than coal, iron and the other staples of the waterways.)

If there had been any lingering doubts about the viability of railways, they were now banished. The railway age had begun. More Acts of Parliament were passed, and the threat to canals began to become painfully obvious. By 1835, parliament had authorized fifty-four new railway lines. In just two years, 1836 and 1837, thirty-nine were waved through.

In 1830 the Leeds and Selby Railway took on the Aire and Calder Navigation. By 1833 the Kennet and Avon was under the cosh, with the founding of Brunel's Great Western Railway, running by 1838. (The company's initials, GWR, were sometimes interpreted, such was the excitement railways generated, as 'God's Wonderful Railway'.)

The Grand Junction Canal also had a rival. The London and Birmingham Railway, designed by Robert Stephenson, son of George, was finished in 1838. The Manchester to Leeds Railway opened in 1839, threatening the traffic of the trans-Pennines lines.

In the 1840s what became known as 'Railway Mania' set in. By 1847 the government had authorized the building of more than 8,000 miles of railway track. (At their peak, in 1840, there were 4,003 miles of canal.)

Gradually the rival lines coalesced into a national network. They were to become organized, amalgamated and efficient. By 1849 the 'Railway King', George Hudson, effectively controlled just under 30 per cent of Britain's railways, the great majority of them owned by just four companies. (Hudson's career ended in tatters; a bankrupt forced into exile.) By 1900, there were 22,000 miles of rail track in Britain.

Every canal in Britain would be adversely affected by

the railways one way or another. Although it's a myth that railways killed the canals stone dead, their story from now on, until the great explosion of leisure on the waterways in recent decades, is one of long, inexorable commercial decline and decay, as many canals fell into disrepair, dried up and rusted and crumbled away. The party was over.

The first victims of the railways were the stagecoaches. Their dispatch was swift. As early as 1843, they had ceased to run at all between London and Bristol. The arrivals and departures of stagecoaches, the great set-pieces of many a Dickens novel, were over. The turnpikes also lost huge amounts of trade, and for a while the roads seemed to be deserted. They were due to come back – with a vengeance. The canals, however, were fighting a losing battle. The tide was unstoppable.

The many disadvantages of canal transport, of which its users had long complained, now came back to bite the canal companies' behinds. It wasn't just that railways were faster, and cheaper to build. Nor that the great strides that had been made in civil engineering, thanks to the canals and roads, meant that the railways could be constructed in lines far straighter than those of the old contour canals of James Brindley. The main problem was inherent in the genesis of the canals. Brindley had dreamed of a whole network, the Grand Cross. But the system, as it stood in 1840, had developed piecemeal, by a series of after-thoughts and add-ons.

Many of the people involved in the early days, though they may have understood the value of a national network, were only really concerned with the stretch of waterway that affected their businesses. The typical canal journey was, and remained, 20 miles. Because of this local perspective there came about the many anomalies of gauge that bedevilled through traffic. Not only narrow versus broad, but the fact that some locks, like those on the Leeds and Liverpool, and the Calder and Hebble Navigation, weren't long enough to accommodate narrow boats. What's more, the old legging tunnels were slow going and caused terrible bottlenecks. Locks too – as any canal enthusiast can testify today – are time-consuming obstacles to negotiate.

Another intractable problem was that the canalsides had, in the densest industrial areas, quickly been colonized by factories, wharves and warehouses. So, short of demolishing these buildings, widening the pounds (as opposed to the locks) in these areas would have been extremely difficult and extremely expensive, if not impossible.

Another tricky problem for canals was the weather. In the winter they would freeze over, and teams of ice-breakers would be set to work. It was not at all unusual for a boat to be iced in for days, and sometimes weeks at a time. Another handicap was the perennial lack of water, especially in summer.

There was little the canal companies could do about gradients, tunnels, locks or the weather. The narrow gauge they were stuck with. But there were things they could, and should, have done – maintaining their canals adequately, for one. Weeds, lack of water, badly neglected towpaths or tunnels, all of these caused delays. So too did silting up, if canals weren't properly dredged. And a low level of available water, whether from silt, or a lack of water supply, or both, meant that boats, if they could get through at all, could carry less tonnage.

As competition from the railways bit, tolls were lowered, but dividends were still paid, meaning that there was less and less money for essential maintenance, let alone improvements.

Even more unforgiveable was the quite remarkable complacency and lack of cooperation between the canal companies where it came to tolls. Not only was the system complex, with different charges for different cargoes, and a range of discounts being offered. Astoundingly, it was impossible to get a quote for a through journey that used more than one canal. Only in 1873 did the government force the canal companies to quote a single toll for through traffic, and this was a cumbersome arrangement whereby the carrying company had to approach every canal on which it proposed to travel. As late as 1906 a carrier told a Royal Commission that he had spent a month to five weeks trying to negotiate a fixed rate over five canals. Four had 'got into the humour of taking it' while a fifth still 'stood out'.

Although railways, in their own mania days, suffered from the same bull-headed short-sightedness and self-defeating competition as did the canals before them, they weren't so stupid or idle as to fail to be able to set up systems that produced a quote for a through journey along several lines. In 1842 a Railway Clearing House was formed, which did just this, as well as administering mutual debts incurred by railway companies and standardizing paperwork. Such a commonsense idea never seems to have occurred to the canal companies.

With huge sighs of relief, many merchants and industrialists transferred their business to the railways. The greed and myopia of the canal owners had done much to help speed them in this direction.

Belatedly, the canal companies themselves began to realize this, and they promptly set about blaming each other. In 1841 the Birmingham Canal Navigations Company, not previously noted for its pursuance of policies promoting the common good, noted that the current malaise, as it affected loss of traffic and the decline in share prices, was 'less to be ascribed to opposing railways, than to the inactivity, want of foresight, and absurd jealousies of the Canal Companies themselves'. Touché.

The Staffordshire and Worcestershire Canal, while doing absolutely nothing to make improvements to its own canal, tried to whip up a united opposition to see the railways off. But the only serious talk of amalgamating the canals came in 1841, when the Oxford Canal, idiotically, was charging exorbitant tolls even though pressure from the railways was becoming critical. Other companies, led by the Grand Junction, threatened to band together to build a new line, which would have cut the Oxford Canal out of the main route from London to the Midlands once and for all. The company lowered its tolls. No further efforts were made to amalgamate canals. Instead, the response of the owners was mostly supine and feeble.

It's true that canal companies did oppose Bills for railways, submitting petitions and publishing pamphlets. But the onrush could not be stemmed. By 1845, only four years after its clarion call, the BCN had concluded that 'it is no longer expedient to carry on any united opposition to the various projected railways'.

In some cases opposition to Railway Bills was merely a

form of extortion. The aim was to encourage, if not blackmail, railway companies to buy the canal companies out, the price including not just the apparatus itself, but also a backstairs promise not to oppose the railways' plans. The Ashby Canal, for example, pimped itself out to the railway company building a line through the Leicestershire coalfields.

In other cases, canal company shareholders, seeing tolls, profits, dividends and share prices all heading south, sold out in desperation, so as to rescue something of their capital, at least, from the flames.

The canals fell quickly. In 1845 the railways bought or leased 5 companies, with 78 miles of canal between them. The next year they snapped up 17, with a combined total of 774 miles of waterway, and the year after that, a further 6, and with them 96 miles of canal.

In just these three years some 80 per cent of the canals that would come under railway control had done so. A fifth of all navigable waterways were now in railway hands. They would end up in command of over a third of the total canal mileage.

The Trent and Mersey sold up to the North Staffordshire Railway Company. The Birmingham Canal Navigations were leased by the LNWR in 1845, and the same company leased the Shropshire Union Canal (which included the Llangollen) two years later. The Kennet and Avon became the property of the Great Western Railway in 1852. Even the well-run and successful Aire and Calder came close to playing footsie with the railways, before deciding to soldier on, as did the Grand Union Canal. The Liverpool and Leeds Canal, having leased its toll receipts, but not the canal itself, to various railways, also stayed in business on its own account, as did the Caledonian.

In some cases where canals were leased, such as with the Birmingham Canal Navigations, the canal companies were able to secure a deal whereby their dividends were guaranteed by the new railway owners. Where outright sales were made, some canals struck a good bargain. In most cases, though, because the companies were negotiating from such a weak position, the sums paid were paltry compared to the receipts they had been raking in before railway competition, and the costs of construction (especially where canals built in mania years were concerned). Financially, the glory days were over.

From the railway companies' point of view, they bought or leased canals partly to prevent opposition to their authorizations, partly to snuff out competition and in some cases with a view to closing the canal and building a railway track instead, either beside the existing canal or in place of it. As leverage, some canal companies pretended that they were about to build a railway of their own. (The Shropshire Union did indeed start building one.) More rarely, a canal was bought or leased because the railway genuinely thought it would enhance its own operation (the BCN being one such example).

After this fire sale, the government – prompted by a lobby led by the Aire and Calder – did take some steps to try to bolster the canals as a going concern. In 1845 the Canal Carriers Act authorized the canal companies to act as carriers themselves. To maximize profits, they were also allowed, thanks to another Act the same year, to vary the tolls on different parts of the canal, which had been forbidden up until this point.

Although the Aire and Calder and, at arm's length, the Trent and Mersey had acted as carriers from their inception, most canal owners had preferred to sit back and rake in the tolls while independent carrying companies did the heavy lifting. After the 1845 Act many companies, most energetically the Grand Junction, turned to carrying themselves. Most discovered they'd been right to give it a miss the first time round, as the trade was arduous and complicated, and profits hard to come by.

The use of fly-boats increased, and even longer hours were demanded of boatmen and canal employees. One of the major cargoes at this time was building materials – for the railways.

The Canal Carriers Act of 1845 had also allowed canal companies, for the first time, to lease other canals. The idea was to try and encourage some much-needed cooperation. Instead, some railways took unscrupulous advantage of this legislation and, through canals they already owned, leased others that were in competition with them – until the government put a stop to it.

It's often said that the railways bought the canals merely to put them out of business. But this isn't quite true.

Railway companies sometimes made noticeable efforts to support some of their canals – just so long as they were in the territory of a rival railway. The Shropshire Union Canal was, for a while at least, supported by the LNWR because it was partly on GWR terrain, while the Huddersfield Broad Canal was neglected, as it was a threat to the railway's own lines.

Where it was commercially viable, which is to say where canals serviced industry, rather than disappearing into a rural cul de sac, the canal system was integrated with both railways and wagonways. Nowhere is this seen to better effect than in the Birmingham Canal Navigations. Here, because of the sheer density of factories on the canals, railways couldn't penetrate, so instead they kept the canals going. Several interchange basins were built, where goods could transfer from one mode of transport to the other. Similar arrangements were put into place in London and Manchester.

But for the most part the future for canals owned by railways was unremittingly grim. The railway companies had no interest in setting up a canal clearing house. And they made sure that tolls on their canals were high enough to make them uncompetitive. (The GWR forced the Kennet and Avon to charge tolls that were 50 per cent higher than the average.) In terms of maintenance and improvements, policies of neglect, inertia and stasis were actively pursued.

Charles Hadfield, the first important post-war historian of the canals, puts it succinctly:

> *Railway companies naturally wished to carry traffic by rail and not by water, since otherwise they would have been competing with themselves. Partly by intention and partly by neglect, the general effect of high tolls, a lack of dredging, closures for leisurely repairs, failure to provide ice breaking, and a general failure to modernise and develop decaying warehouses and wharves, failure to provide or maintain cranes and no effort to obtain business served to divert trade from the water to the land.*

But the railway companies couldn't simply trash their new purchases. The government, waking up to the danger posed by the elimination of the canal network, and the resulting monopoly that would be enjoyed by the railways, stipulated in 1873 that the canals must be 'at all times kept open and navigable for the use of all persons desirous to use and navigate the same without any unnecessary hindrance, interruption or delay'.

Although this decree was to be honoured more in the breach than the observance, the railways had to keep the canals flowing one way or another. A suitably Victorian 'Act of Abandonment' was needed to close a canal down.

The system stumbled on. The fly-boat trade in expensive or perishable goods that were needed to be moved quickly was hit hard by the railways, although it continued. Canals, from the 1840s until the end of their commercial traffic in recent times, mostly carried what had always been their staple cargoes: high-volume, low-unit-cost items such as coal, iron, steel, limestone, sand, building materials and fertilizer. But while the amount of tonnage was similar to what it had been before, and in some cases even increased as the century progressed, overall toll receipts declined.

Steam

Attempts were made to speed the traffic up. One successful innovation was the introduction of steam tugs. A pioneer here was an American inventor, Robert Fulton, who built steam boats for the Duke of Bridgewater in the late 1790s. A steam trial was also held on the Sankey Canal in 1797. In 1803 the Forth and Clyde Canal introduced the steam tug *Charlotte Dundas*, designed by William Symington. But the wash from the boat damaged the banks, and the experiment was discontinued.

Steam tugs found more favour on river navigations, where wash wasn't such an issue. On the Severn between Gloucester and Worcester, for example, sometimes as many as twenty-four narrow boats, in two parallel lines, were towed by tugs. But locks, unless they were of huge size, were a considerable problem, as of course a string of boats had to be uncoupled and then manhandled through them. Tugs were generally to be found on longer pounds.

They became more economical after the adoption of propellers in place of paddles, in the 1850s. Canal companies, though, were not enthusiastic, as the wash created by the faster speed of the boats carried even more damage to the banks. In 1859 the Ashby Canal Company banned them altogether. After a court case, in which their effects were carefully assessed, steam boats were allowed, so long as they kept to a limit of 4 mph. Other canals followed the Ashby's example, including the Leeds and Liverpool and Grand Junction Canals.

The always go-ahead Aire and Calder had made extensive use of tugs, beginning in the 1830s. By 1855 two-thirds of its traffic was steam-hauled, and the service was offered free to 'bye-traders', or independents. The famous 'Tom Puddings' appeared in 1865. Its use of steam – and, crucially, its adaptation of its canals to accommodate steam boats – was a major reason that the Aire and Calder survived as well as it did.

One of the most useful functions of steam tugs was to

haul boats through tunnels (thus putting the disgruntled 'leggers' out of business). They were also used for ice-breaking and routine maintenance work, such as dredging.

There were sporadic attempts, beginning in the 1820s, to introduce steam engines into the canal boats themselves. This didn't really take off until the 1860s and was never truly successful. For one thing they needed an extra crew member to operate them. And an insuperable problem was the great amount of precious cargo space they took up – a particular drawback, of course, in the case of narrow boats.

What's more, the fact that steam-driven boats were faster than horse-drawn ones wasn't of much use on narrow canals. There is a limit to how fast a boat can travel, which depends on the width and depth of the canal. Too fast, and not only does the wash damage the banks, but water heaps up at the fore-end, making the boat more difficult to steer. Steam-powered boats were more often to be found on the broad sections of the canals. The carrying firm of Fellows, Morton and Clayton was an early adopter and built several narrow boats with steam engines. They were paired with engine-less boats or 'butties'.

The development of steam power increased coastal traffic from port to port, at the expense of canals in the interior (the Caledonian being a special case in point). However, the increase in shipping handled by ports was good news for those wide canals and navigations which directly served them, such as the Aire and Calder, the Gloucester and Sharpness Canal, and the Weaver Navigation. Canals that linked to estuaries were the ones that best weathered the storm.

The rise in coastal traffic was a disaster for the agricultural canals, never very successful in the first place, as more and more grain was imported from the 1870s, in ever-larger ships.

By now it had become apparent that the railways had snapped up a virtual monopoly of inland transport. This was demonstrated by the excessive rise in their charges.

What's more, imported goods were beginning to be seen in Britain in greater numbers than ever before. There was a perception that the railway companies were offering foreign firms reduced rates to use their lines. It thus became a matter of urgency to reduce the cost of carrying British freight. Economists and politicians had also woken up to the fact that waterways abroad were controlled, and in many cases subsidized, by their governments.

Attention now turned to the canals. Could these not be shaken up so as to provide some competition, and lower prices? Clearly those owned or leased by railway companies were hardly going to be allowed to undercut their masters. But what about the independent canals, amounting still to two-thirds of the total mileage? Could they not be revamped, to everyone's benefit?

Not for the last time, grand schemes were proposed. To widen the narrow canal network, for example – still commercially important despite its drawbacks.

In the 1880s serious consideration was given to taking the whole network under government control – if not nationalizing it altogether – in order to bring such measures into effect. Clearly, private enterprise alone would not be capable of funding the scale of work needed.

Construction workers on the Manchester Ship Canal, *c.* 1890

The construction of boat lifts and inclined planes to replace the troublesome locks was much discussed. But though there was plenty of talk, little action was taken. The famous Anderton Boat Lift and later the inclined plane at Foxton, on the Old Grand Union, were built, but proved to be isolated examples.

There was, though, one very notable success. Just as the Bridgewater Canal had come into being because a river navigation was being too greedy, and then a railway had been built because the waterways were charging exorbitant fees, so a canal was built because a railway had been taking too great an advantage of *its* monopoly. The only major canal built after the great period of construction ended in the 1840s was, significantly, a leviathan: a ship canal.

In 1872 the Bridgewater Canal was bought by a railway consortium, thus giving railways a monopoly of the transport between Manchester and Liverpool. Four years later the Manchester Ship Canal was conceived as a response to this. An economic depression at the time, along with trenchant opposition from railway interests, forestalled any action, but the idea was picked up again in 1882. An Act was passed three years later, but finding the money proved difficult. Not until 1894 was the canal ready for business. It was officially opened by Queen Victoria herself.

The Manchester Ship Canal took almost the same route (although more directly) as Brindley's Bridgewater Canal, the waterway that had set canal building in motion in the first place. By a pleasing congruency a descendant of the Duke of Bridgewater was installed as chairman. The last part of the canal to be built was the swing aqueduct at Barton, which carries the Bridgewater Canal over the new one. The old Barton Aqueduct – the marvel that had made its name when it had, in its own turn, straddled the River Irwell more than 130 years earlier – was demolished.

After an unsteady start the ship canal flourished. Its 36 miles became a de facto dock, linked – by road – to factories that had set up shop along its banks, including those of Trafford Park, at one point the largest industrial complex in the world.

Other ship canals – including one from Birmingham to Liverpool – had been proposed during this period of pondering on '*what can be done?*' But the costs were prohibitive, even if the ideas made commercial sense. The examples of the Gloucester and Sharpness Canal, the Manchester Ship Canal and the Aire and Calder Navigation from Ferrybridge to Goole offer a tantalizing example of what could have been done to prolong the commercial life of the inland waterways. Size mattered.

A steamer enters Irlam Lock in Manchester

Diesel

In the years leading up to the First World War diesel engines began to be adopted and were to transform the canals. In 1906 experiments were first made with a gas suction engine. In 1911 a semi-diesel engine was successfully tried out, and the next year Fellows, Morton and Clayton built the motor boat *Linda*. It used a Swedish Bolinder engine, which became one of the commonest makes on the canals. (Fellows, Morton and Clayton boats were known as 'joshers', after Joshua Fellows, one of the founders of the company back in 1889.)

The diesel engines were smaller and more reliable than steam-powered ones, and, although horses still towed boats up until the Second World War (and, in rare cases, right up until the 1960s), diesel increasingly took over. The team of a motor boat – sometimes called a 'monkey' boat (to the displeasure of the boaters) – and butty was to become the norm on the route between London and the Midlands. (On broad canals an engine-less boat was known as a 'dumb barge'.) This arrangement maximized not only tonnage but also living accommodation.

Fellows, Morton and Clayton was the biggest carrier on the canal network, with over thirty depots on the Grand Junction, Trent and Mersey, Staffs and Worcs, Coventry, Oxford, Birmingham and Nottingham Canals. Thus it had plenty at stake where it came to keeping the canal network alive.

In 1904 the then-chairman of the company published the most important book ever written about British canals, a last-ditch attempt to boost their commercial viability. This was the indefatigable, Eton-educated figure of Count Henry Rodolph de Salis, who, beginning in 1887, spent eleven years travelling in his steam launch, exploring every yard of the 11,000 miles of inland waterways then navigable. It is, of course, a feat that will never be repeated. He did so to produce *Bradshaw's Canals and Navigable Rivers of England and Wales*. It was a companion to the celebrated railway guides that were first produced in 1839.

The aim was to provide – for the first time ever – a comprehensive gazetteer of the waterways, with information about the dimensions of the locks, bridges, tunnels and pounds, as well as the 'contact details' for the company concerned.

A passage in the book's preface should have been stapled to the forehead of every single person involved in canal administration.

Whatever may be said on the question of the unfair starving and stifling of canals by railways, there is no doubt that in the first place the canals had largely to thank themselves for it. In many cases the canal companies forced the railways to purchase or lease their undertakings at substantial prices before constructing their lines.

Canals in their day reached a far greater pitch of prosperity than the railways have ever attained to, but they suffered fatally, and do so now, from the want of any serious movement towards their becoming a united system of communication. Each navigation was constructed purely as a local concern, and the gauge of locks and depth of water was generally decided by local circumstances or the fancy of the constructors without any regard for uniformity. The same ideas of exclusiveness appear to have become perpetuated in the system of canal management; there is no Canal Clearing House, and with few exceptions every boat owner has to deal separately with the management of every navigation over which he trades.

De Salis points out that only in 1897 were quotes for through tolls readily available on the main route between London and Birmingham, then controlled by four separate companies. In a plea that was to be echoed down the next six decades, he remarks: 'Even in the present imperfect state of the system as a whole, many tons of traffic might with advantage be conveyed thereon which are now carried by railway at increased cost.'

Two years after de Salis's book was published, a Royal Commission was set up to look at the whole transport network. It sat for five years and, in twelve volumes, reported on every aspect of the canals, while considering their future – if they had one. Its conclusion, bitterly resisted by the railway companies, was that the canal network, even though it had been treading water since the 1850s and had not shared in the transport boom of the late nineteenth century at all, was nevertheless viable and of vital strategic importance and should be revived. It recommended sweeping and wholesale improvements to the Grand Cross, centred on Birmingham, replacing locks with lifts or inclined planes.

The scheme would have been extremely expensive and was quietly forgotten. The last chance to make a real, systemic difference to the canal network was thus lost. Whether, had it been undertaken, it would have rescued the canals' commercial future is one of the great 'what ifs?' of British economic history.

During the First World War all transport was placed under the control of the government, except the canals, which declined further as boatmen joined up or sought better paid work in munitions factories and in road transport. In 1917, fearing that the whole system would collapse, the government took it under its wing. To try and arrest the alarming decline in the numbers of boatmen, in 1919 wages were raised by a third (to the fury of the carrying companies, when the canals were returned to their former owners later that year).

During the war years, a third of the canals' total tonnage had been lost. It was never to come back. Nor were many of the boaters who'd left for dry land. Road haulage, on the other hand, by an unhappy symmetry, had tripled.

The introduction of the eight-hour working day also damaged the canal companies' economic viability after the war (in so far as an industry that relies on long working hours to make up the shortfall in its commercial sustainability can be said to be truly viable). Although the unionized canal workers on land benefitted, the boatmen (and their families), still paid by the trip, 'worked eighteenth-century hours, for nineteenth-century wages, well into the twentieth century', as a later carrying company owner neatly put it.

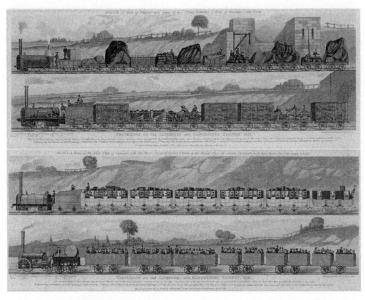

Two freight and two passenger trains on the Liverpool
to Manchester Railway

Some boatmen joined trade unions, but this was rare. Strikes, however, were not unknown. In 1923 there was one at Fellows, Morton and Clayton that lasted fourteen weeks. There were others, dating back to the end of the previous century.

Between the wars canals under railway ownership were allowed, indeed encouraged, to fall into desuetude, and were even more unfairly starved and stifled, to use de Salis's words. The independent canals too lost traffic. While the vital London to Birmingham route kept going, as did the estuary waterways, canals in the interior – some of which had always been peripheral, or uneconomic, or both – limped into oblivion. Casualties of this inter-war period of attrition included practically all the canals of south Wales, and the Ashton, Droitwich, Grantham, Huddersfield Narrow, and Thames and Medway Canals. By 1939 traffic on the Kennet and Avon had virtually ceased.

Only those canals with the very strongest commercial imperative weathered the storm. For now.

Life on the Cut

In the early days of canals boatmen were drawn from the local area, as long trips were comparatively rare. As were cabins on canal boats. Once the competition increased, and as more valuable goods needed to be moved more quickly, longer journeys became common, and the boatmen began, gradually, to live on board the boats.

Before this step change in the 1790s, complaints about boatmen's behaviour, though they were made, were fairly sporadic. Men who worked the day boats had a reputation to uphold in their local community. But as journeys grew longer, and the men – especially the young and unmarried ones – became more and more cut off from straight society, their behaviour, as a class, grew more disreputable. After the end of the Napoleonic Wars the reputation of canal boatmen began to worsen.

Pilfering on the canals became rife. Even by the end of the eighteenth century many boatmen had come to see it is an inalienable right to skim off some of the coal, which they could then sell to private customers up and down the canal. So widespread was this practice that collieries, notoriously wasteful of their coal, added up to a tenth more cargo in the expectation that some of it would be purloined. Later, when journeys became longer, and coal was needed for cooking, the idea that free fuel was a perk of the job seems fair enough.

Unscrupulous boatmen – always, of course, in a minority – also helped themselves to cloth, wool, sugar – anything that wasn't nailed down and could be sold en route. Receivers of stolen goods set up depots near the canals. The thefts were difficult to detect, as often only quite small quantities were stolen. Canal water or some other useless substance was added to the cargo to make up the weight of whatever had been salted away. Warehousemen were sometimes in on the deal, bribed to look the other way if a cargo had obviously been tampered with. By the time the goods reached their destination – and it might be overseas – they would usually have passed through so many hands as to make the identity of the culprits hard to pin down. Of course, all these small thefts added up to a handy profit for the thieves and a considerable loss for the cargo owner.

The stealing of more expensive goods, such as china, silk or alcohol, was clearly far more of a problem. In 1795 the Trent and Mersey Company announced that wines and liquors would be kept under lock and key while in transit. But many ingenious methods were found to take a bite out of such valuable cargo while preserving its unblemished appearance. Some dedicated thieves among the boatmen worked out a way to duplicate the seals that had been placed on the higher-priced items. If barrels of drink were on board, the standard method to sample their contents was to slide the metal hoops to one side and drill a small hole into the cask. The beverage could then be drawn off, replaced with water and the hoop put back in place.

Raiding the drinks cabinet didn't always go according to plan. On one occasion a crew carrying kegs of rum, brandy and gunpowder decided to open one of the barrels of liquor. Already in the grip of the grog, they were too impatient for the time-honoured method of carefully boring a hole behind one of the hoops. Instead, in their befuddled state, two men took an axe to one of the barrels. Harry Hanson, in *The Canal Boatmen*, takes up the story: 'Their lantern's flickering glow was inadequate as an illuminator, but successful as a detonator, since they chose the wrong barrel, blew up themselves and the boat, and set fire to three haystacks.'

The criminal fraternity among the boatmen also stole from other boats, or indeed from their own crewmates, which meant that honest boatmen needed to be constantly on their guard, for both their own and their employers' sakes.

Another worry for the carrying companies, especially on longer trips, was that they had to entrust the boat captain with considerable sums of cash to pay the tolls, and thus there was a concern that he might abscond with the money.

Boaters were notorious for poaching and often kept dogs on board for this purpose. The canals snaked through the estates of many rich landowners, who had, with the help of Enclosure Acts, taken over what had been common woodland and turned it into their own private game reserves. Penalties for poaching were draconian. Transportation for seven years could be the price of feeding your family if the harvest had failed or wages were in short supply. The boaters, able to hop on and off their boats at the dead of night, were perfectly placed to take advantage of any game that might be had. Even deer. Turnips, vetches, fish, hens, geese and sheep were also carried away. Clover for the horse would be scythed from a landowner's field, and his cows surreptitiously milked. All boatmen, according to the head of Pickfords, 'were to a greater or lesser extent, poachers'.

Boats had, of course, no toilet or washing facilities on board, even when cabins started to be fitted. Calls of nature were answered by the simple expedient of nipping behind a convenient hedge. In urban areas, it isn't hard to imagine the unholy stench along the towpaths, especially in the densely concentrated canalsides of Birmingham, where chloride had to be used on a regular basis to try to freshen things up.

Horses were the lifeblood of the canals and yet they were routinely maltreated. They were often worked fourteen hours a day, half-starved and savagely beaten. In 1783 one such was described as being

> *the skeleton of a horse, covered with skin: whether he subsists upon the scent of the water, is a doubt: but whether his life is a scene of affliction, is not; for the unfeeling driver has no employment but to whip him from one end of the canal to the other.*

Cases of horses being worked or beaten to death and then thrown into the canal were by no means rare. Sometimes they simply fell into the water and were drowned thanks to sheer exhaustion. As more cuts were made into the main lines of canals, there were instances of badly constructed, rickety bridges collapsing under the weight of the horse and drowning the unfortunate animal in the canal below.

More rarely boats were towed by mules, or by pairs of donkeys (always referred to simply as 'animals' by their owners) as on the Stroudwater, Worcester and Birmingham, and Shropshire Canals.

In 1822 the first ever legislation in the world aimed at stopping cruelty to animals was passed. And yet in 1841 it was still being said that very few beasts were 'treated more cruelly than the Boat Horse'.

None of this did the canals' reputation much good. Nor did the undoubted fact of drunkenness amongst some canal crews. And sometimes amongst lock- and toll-keepers, who could be bribed and plied with drink so as to encourage them to turn a blind eye when required.

Canal leggers on the Trent and Mersey Canal

One reason for the propensity for boat crews to drink, though, was that fresh water was almost unobtainable on the canals until the very end of the nineteenth century. The many canalside pubs solved the problem of slaking thirst, and what scanty references to the details of the boaters' drinking habits we have tend to show that copious draughts of weak beer were the favoured tipple.

Leggers, too, had gained a bad reputation, hanging around at either end of the tunnels, being generally abusive and trying to extort money for their services. In 1825 officers of the Grand Junction Canal received a letter complaining of 'the nuisance arising from the notoriously bad characters of the persons who frequent the neighbourhood of the Tunnels upon the plea of assisting Boats through them'. Eventually the canals licensed the leggers, who now wore special identifying insignia.

Some canal carriers seem to have had a particularly jaundiced view of the men in their employ. 'I do not know that there is any class of men in this country that is so bad,' a carrying company clerk remarked. A warehouse owner added that 'they have neither fear of God or man'.

A disgraceful incident that quite properly caused widespread anger and disgust took place in 1839. Mrs Christine Collins engaged Pickfords to transport her from Liverpool to join her husband in London. At dawn on 16 June she joined a fly-boat on the Trent and Mersey Canal at Preston Brook. (Fly-boats rarely carried passengers.)

There were four crew members on board: three men and a twelve-year-old cabin boy. The men were known among their fellow boatmen as notorious ruffians and drunkards. Still groggy at the start of the journey from the excesses of the night before, they began drinking heavily the same evening, helping themselves to the cargo of rum that was also aboard. Some time in the early hours of the next morning, near Rugeley, the men attacked their passenger, raped her, murdered her and threw her dead body into the canal. The two main perpetrators of this bestial crime were hanged. The third man was transported for life because he couldn't definitively be placed on board when the murder took place. (As he was the first to confess, another theory is that he 'copped a plea' in return for escaping the noose.)

What especially dismayed the court, and the public who read about the crime in their newspapers, was that Mrs Collins had on two separate occasions expressed her growing nervousness about her attackers – as they turned out to be. At Stoke-on-Trent she complained to the Pickfords office about the bad language of the men, and that the captain, James Owen, was paying unwarranted attention to her. She asked to complete her journey by stagecoach. That proving to be full, she was encouraged to get back on board the boat. Arriving at Stone – the Trent and Mersey Company's head office – at the dead of night, she complained again, this time to a company clerk. But he, frightened of the notorious temper of the captain, declined to help her, merely advising her to make a report to Pickfords at the journey's end.

Later the same night, at 5 a.m., a lock-keeper and his wife heard the screams of a woman coming from inside the boat. But Owen was able to convince them that it was his wife, also drunk, whom they had heard. They saw a woman being bundled into the cabin but, somewhat incredibly, allowed the boat to continue on its way. By the time another lock-keeper, tipped off by the cabin boy that something was badly amiss, and aware of Owen's bad character, informed the police, it was too late. Mrs Collins was already dead. No one had stepped forward to protect her, nor lift a finger to try to prevent the crime that eventually took place.

After this shameful incident, both the scrutiny of, and suspicion of canal boatmen increased. An invaluable, albeit *parti-pris* insight into the seedier side of life on the canals is provided by the proceedings of the Select Committee on Sunday Trading on Canals, which was convened in 1841, while the memory of Mrs Collins's murder was still fresh. The committee had been set up to investigate (and really to prove) the widespread belief that, because boaters worked on Sundays, this in itself impelled them towards immoral behaviour. (An earlier committee, on the Observance of the Sabbath, had sat in 1832 and had also inquired into the boatmen's religious habits. This was at the time of a religious revival in Britain.)

Various reformed boatmen-cum-thieves, and fences, divulged the different dodges that dishonest boaters used to help themselves to their cargo and otherwise pull a fast one.

Sir George Chetwynd, a former chairman of the Trent and Mersey Canal Company, told the committee that he had spoken to William Ellis, the boatman who had been involved in the murder of Christine Collins but spared execution: 'Since he has been in Stafford Gaol he has made such Disclosures as beggar all Description of the Abominations that are committed on Canals, and the Habits of Boatmen, their pilfering, and Habits of horrid Depravity.' Asking him what could be done to 'better the Condition of Boatmen', Chetwynd received the very satisfactory answer: 'putting a stop to the Sunday Trading'. The Select Committee failed to achieve this. (Although some carrying companies did insist that their employees did not work on Sundays.)

By the time these committees were convened, a life on the cut had become ever more unattractive. The skippers made a reasonably good living (although they earned it), and it was the norm for a permanent hand eventually to become a captain in his turn. But casual hands, hired by the trip, were notoriously badly paid. And for everyone the hours were long and arduous.

'Respectable' working men shunned the boats, and instead the canals had to offer bed and board to ruffians, petty criminals and all-round ne'er-do-wells. It became the custom for such desperadoes to hang around at the major junctions, such as Braunston, seeking casual work. A witness to the 1841 Select Committee admitted that 'I spent my time lurking in fields where game lay, sometimes in beershops, public houses, and bawdy houses. When not in honest employ I was maintained by poaching and stealing.'

The committee was particularly concerned about the moral welfare of teenage boys working the canals. Another witness reckoned:

it is generally such Boys as those who have run away from their Parents, or committed some improper Act, and they come to the Banks of the Canal, and there they are sure of getting Employment; and frequently these Boys have robbed their Masters and then run away from them.

It's easy to see why this depressing litany of wrong-doing gave outsiders a lastingly jaundiced view of canal boatmen.

But, of course, good deeds are less publicized than bad ones. The fact that the canal system worked at all gives the lie to any notion that boatmen were routinely dishonest. Had boat captains regularly made off with the cash entrusted to them, the whole system would have sunk without trace. That pilfering went on we know to be a fact because of the evidence given to various committees by the nogoodniks who had been doing it. But it must be the case that such men were in a small minority. Carriers rarely complained of pilfering – and where they did it was at inland ports open to outsiders. Unless we are to believe that large-scale theft was undertaken so cleverly that these sharp businessmen simply didn't notice they were coming out on the short end, the conclusion must be that the vast majority of boatmen were either perfectly honest or, at least, honest to a very considerable degree.

Nor is it likely that, to a man, they were routinely foul-mouthed, drunk or prone to violence. Unless sorely pressed by obstacles strewn in their paths, the majority were perfectly well-behaved and honourable men. 'That their manners and deportment in general are rough and unpolished everyone will admit,' wrote an anonymous correspondent in the letters page of a newspaper, 'but morality and good manners are not synonymous terms, and experience has shown that a person of rough exterior is quite as likely to possess an honest heart as one whose demeanour is more polite and whose manners are better polished.'

What's more, even though they continued to work on Sundays, from the 1840s onwards the behaviour of boatmen seems to have improved, and fewer complaints against them were made. That this should be so might have been put down, by many a Victorian moralist, to the civilizing influence of a good woman. Although boys continued to run away from their masters to work on the floating sweatshops of the fly-boats, a whole new class of employee had begun to appear on the slower boats. The fairer sex had begun to work the canals.

When they began to do so, and in what numbers, is frustratingly unclear, as is so much of the history of the boaters. Until recently it was thought that the economic pressure imposed by railway competition pushed women on to the boats in the 1840s. But further research, by Harry Hanson and others, indicates that they began living on the boats in the early years of the nineteenth century. By 1832, when the Select Committee on the Observance of the Sabbath was sitting, it's clear that their presence was well established.

There were several reasons why women and children began to live on and work the boats. The first was simply that it was economically advantageous. The fee for a single journey was paid to the captain of the boat. From this money he had to pay his crew. If his wife and children *were* his crew, savings could obviously be made. (Of course it was perfectly common at this time for whole families to work together – as in weaving, for example.)

Another impetus for wives and children to live on board was the increasing length of the journeys: otherwise the boatman wouldn't see his family for weeks on end. (This continued to be the case for the fly-boat crews.)

There was perhaps yet another reason why the boatmen's wives were keen to accompany them. It was by no means unknown for a boatman to hire a female companion to share his journey on a long voyage (a suggestion made, inevitably, at the Sunday Trading witch-hunt). But another possibility is the more straightforward and prosaic one: that the husbands and wives were no less likely than those on land to love each other, desire each other's company and not want to be parted from their children.

The slump that followed the Napoleonic Wars seems to have led to an increase in the number of 'family boats'. Maintaining a house on land when you could accommodate your family on board made little financial sense, even if it was highly attractive in terms of more civilized living conditions. Nevertheless, especially in more prosperous periods, many boating families did keep their 'land houses'. Sometimes there was a mix and match – where possible the family lived in their house on land and took to the boats only on long journeys, or in the summer months.

Although family boats were present on broad canals, such as the Leeds and Liverpool, and Rochdale Canals, they were to be found in greater numbers on the narrow canals, particularly in the Midlands and north-west. This meant that both adults, and as many children as could be crammed in, lived in a cabin typically measuring less than 9 feet long, 7 feet wide and 6 feet tall. Here they would cook, eat, sleep, make love, give birth, fall ill and die.

However, it is not the case that the families always slept on board, even on long journeys. Horses could hardly bed down on narrow boats, and so the many canalside inns, like their counterparts on the public highways, offered stabling. The families often stayed the night at these pubs.

The boat families became self-sufficient, and self-perpetuating. Boys – and girls – could be steering the boats by the age of nine.

Like any body of men and women living apart from the mainstream, they tended to be regarded with suspicion, distrust and sometimes distaste. Their cold-shouldering by the world on land made the boaters turn in on their own resources. They became a people apart.

In the second half of the nineteenth century the lives of the boaters gradually, if still hazily, came into focus. In 1858 Charles Dickens's famous weekly magazine, *Household Words*, ran John Hollingshead's article 'On the Canal', a description of a voyage on a fly-boat, the *Stourport*, that the author had undertaken from London to Birmingham.

Hollingshead, like Dickens himself, had a sharp reporter's eye for detail, and his short account is an invaluable record of the canals and the men (in this case, not women, as it was a fly-boat) whose lives depended on them. He describes his companions as wearing 'short fustian trousers, heavy boots, red plush jackets. Waistcoats with pearl buttons and gay silk handkerchiefs slung loosely around their necks,' along with distinctive blue stockings and a sailor's cap. The *Stourport*'s skipper was Captain Randle. He had spent fifty years of his life on the canals, man and boy. He had a land home in Stoke, but such was his workload that he only got to see it three times a year.

Hollingshead, naturally, knows all about the bad reputation the boatmen had won for themselves. But he is impressed with the extreme tidiness of the vessel. What's more, he found that:

Beer and spirits were little used, and a pipe being a rare indulgence. Melancholy pictures of drunken brawls, improper language, constant fights, danger to life and property, hordes of licensed ruffians beyond the pale of law and order, which my cheerful friends had drawn the moment they heard of my intention to make an unprotected barge journey, all proved false before the experience of a few hours, and shamefully false before the further experience of a few days. We were inmates of a new home, and friends of a new family; whose members were honest, industrious, simple, and natural – too independent to stoop to the meanness of masquerading in adopted habits and manners with a view of misleading the judgment of their guests.

The real interest in the piece is the character of the captain himself – despite the fact that he is presented as something of a buffoon throughout. He possesses, we are told, a 'not over keen intellect'. Randle himself admits he is no 'scollard': 'I am a doonce, an' I knows it; but I made my boy larn to read an' write, an' if I could affoord it, he shouldn't be 'ere now.' Hollingshead cheerfully assures him that he hasn't lost all that much by not being able to read. This is the first of many examples of well-to-do outsiders airily dismissing boaters' own frustrations with their illiteracy. What is striking is how strongly Randle disagrees: 'This be a hard life in winter time, an' I'd be well out of it at my age if I'd been a scollard.'

The tone of the piece changes completely when Hollingshead describes the family boats, with

children varying in number from two to ten, and in ages from three weeks to twelve years. The youngest of these helpless little ones, dirty, ragged, and stunted in growth, are confined in the close recesses of the cabin . . . rolling helplessly upon the floor, within a few inches of a fierce fire and a steaming kettle . . . fretful for want of room,

air, and amusement . . . sickly, even under their sun-burnt skins;
waiting wearily for the time when their little limbs will be strong
enough to trot along the towing-path; or dropping suddenly over the
gaudy side of the boat, quietly into the open, hungry arms of death.

This Dantesque perception of the family boat as a floating shop of horrors was increasingly to take hold. In 1875 a moral feeding frenzy was kicked off by the publication of a book called *Our Canal Population*. The revelations it contained – or purported to contain – were deliciously shocking for Victorian moralists and those interested in sleazy goings on – or both.

The author was a remarkable individual. Born into poverty, George Smith (or 'George Smith of Coalville', as he invariably styled himself) had become the well-paid manager of a tile-making company until he decided to write a book exposing the brutal conditions suffered by children as young as eight working in the brickyards. (Lifting the lid on this was to make him jobless, and extremely poor, for the rest of his life, 'blacked' by potential employers.) Smith, by this stage believing he was on a mission from God, became known as 'The Children's Friend'.

In the early 1870s he turned his attention to life on the canals. And he was appalled at what he saw. The thrust of his subsequent book was meant to be sympathetic, arguing for the betterment of conditions on the canal for children.

Overcrowding on the boats was his main concern. It was, after all, not unknown for families of up to ten to share a cabin little more than fifty square feet – though this was rare, and families of five and six were far more common. But certainly overcrowding was a serious issue.

(As it was on land, where several slum families often occupied the same fetid basement.)

For Smith, cleanliness was close to godliness:

Their habits are filthy and disgusting beyond conception. I have frequently seen women in a half nude state washing over the side of the boat as it was moving along, out of the water of the canal, upon the top of which had been floating all manner of filth. They wash their clothes – those that do wash – out of the canal water and instead of their being white, or near to it, they look as if they had been drawn through a mud hole, wrung and hung out upon the boat line to dry.

(With no laundry facilities on the canal, and with lock-keeper's wives expressly forbidden, in many cases, to take in washing for money, it's hard to see what alternative these women had.)

Into what was already a rich stew of moral disapproval, the fevered brain of George Smith stirred in a strong dash of sex. According to his sensational book, canal boats were little better than a flotilla sent from the Cities of the Plains. Not only were irreligion, drunkenness, violence and promiscuity rife. Incest was also on the menu. Children were, he said, 'living and floating on our canals and rivers in a state of wretchedness, misery, immorality, cruelty and evil training that carries peril with it' in 'hot, damp, close, stuffy, buggy, filthy and stinking holes, commonly called boat cabins . . . In these places girls of 17 give birth to children, the fathers of which are members of their family.'

Smith was able to add surprisingly accurate figures to his charge sheet. Of the 100,000 people he claimed were living on boats, 90 per cent were drunkards (and foul-mouthed ones at that: 'Swearing, blasphemy and oaths are their common conversation'). Only 2 per cent were Christians, while 60 per cent of the couples were living in sin.

As with his whole book, Smith was exaggerating. Although subsequent historians haven't been able to work out accurate figures with the same remarkable ease that he did, it's since been estimated that somewhere around 27,000 people slept on boats, and of these perhaps only around 7,000 to 9,000 were families living permanently on board. And even here it may have been that they lived on the boats only part of the time.

Although hard evidence is hard to come by, it would seem, certainly by the turn of the century, that though they couldn't attend church on a Sunday, most boaters were christened and married in church. And died and were buried as Christians. As for the demon drink, Smith is hard to refute completely (despite what Hollingshead said), as other, more qualified witnesses confirmed that there was much inebriation amongst the canal population, at a time when it was becoming less common on land.

As for the common perception that the boats were filthy, of course some were – and remained so until the traffic died away in the 1960s. But these seem to have been a very small minority. As one informed observer put it, 'The cabins on the whole are kept fairly clean, some remarkably so, and among some of the younger and steady boat men, there seems to be an emulation to make their cabins as clean and smart as possible.'

What Smith was particularly exercised about was the perception that older children were at risk by sharing a cabin with adults. Incest on the boats was not unknown, of course. Nor was it on land. But it was rare. What particulary concerned Smith and other outsiders was what they saw as the moral danger of canal children being routinely exposed to their parents' sexual activity and thus morally corrupted. What became known as 'indecent occupation' became a hot topic.

There were other aspects of children's lives on the canals that troubled observers. One was the practice of parents giving children away to strangers. Ironically, of course, this might have happened precisely because of overcrowding. But there were sad cases where children were sorely abused by their new owners. In one revolting incident an eight-year-old was worked, and beaten, to death.

But, as usual where moral panics about the canals were concerned, the frequency of children being given to strangers was grossly exaggerated. More common was the surely more reasonable practice of children being sent to live on boats owned by relatives, especially grandparents, where there was more room.

The result of Smith's agitation was the Canal Boats Act of 1877, which required all boats to be registered, set a limit on the number of people who could legally live in one cabin (parents and three, later two, children) and dictated that girls over twelve, and boys over fourteen, were not allowed to share a cabin with a man and his wife. (Incidentally making marriage in effect a legal requirement for boat couples.)

Local authorities were ordered to inspect the canal boats that passed through their districts. In practice the Act proved to be next to useless as it prescribed few penalties for infringement. In 1884, after more campaigning on the part of Smith and his supporters, a stronger Act was passed. A canals inspector was appointed, and the regulations were more assiduously enforced.

This had an unforeseen benefit for the boaters. Through the inspectors they were now able discreetly to put pressure on the boat owners to make improvements in their craft and maintain them to a higher standard. From now on the great majority of canal boats were to be clean and a little more comfortable, and social behaviour, unsurprisingly, improved, while excessive drinking gradually declined.

Children in some cases now slept in a makeshift cabin at the fore-end of the boat, which was far from ideal because of damp. More often, they had to leave the boat when they became too many in number. Thus the proscription against overcrowding, while it brought much-needed benefits, also led to boaters leaving the canals. It was common for girls over the age of twelve to be sent into domestic service. Boys, if they couldn't (or didn't want to) find a berth on another boat, left for a life on land. By 1892 the first canal inspector reckoned that only 1,000 boats were occupied by families on a permanent basis.

Where George Smith and the other moralists who criticized the lives of the boaters were undoubtedly on firmer ground was in the matter of illiteracy. Although the figure he gave of 95 per cent may be too high, there is no doubt that by far the great majority of the boat population were unable to read and write. This was commented on by practically everyone involved in canal administration.

The Canal Boats Acts of 1877 and 1884 had attempted to redress the situation and encourage the boaters to send their children to school. But because of their peripatetic lifestyles – and the fact that they were often on the move from dawn to dusk in order to make a living – attendance at school was extremely difficult to arrange. Illiteracy remained an intractable problem among the boatmen and their families well into the 1940s, if not beyond.

Not all the dry-landers who took an interest in the moral welfare of the boaters were quite as censorious as George Smith of Coalville. The people who ran missions, such as the Incorporated Seamen and Boatmen's Friend Society, formed in 1846, were well liked by the canal folk, as they concentrated on helping them rather than finger-wagging. They also assisted with reading and writing letters. Starting in 1841, several canalside chapels were built (several of which were afloat), which also doubled as Sunday Schools. A school exclusively for boat children was founded, in Brentford, in 1896. Another was set up at Paddington shortly afterwards. By 1910 there were thirteen such schools around the country.

Children at a barge school at West Drayton, 1933

Chapels were sometimes paid for by the canal companies. But they were less keen to dig into their pockets when it came to providing sanitation. Lavatories were in short supply and generally in such a disgusting state that no one used them. There were few washing or laundry facilities at the basins, depots and wharves. Fresh water supplies were also thin on the ground until the 1890s. Clearly the boaters' moral welfare was more important to the canal companies, and Victorian moralists, than their physical well-being.

After the First World War, the pattern of life on the cut seemed to settle back into its long-unchanging traditions. At yet another inquiry into the family boats, a manager at Fellows, Morton and Clayton was asked, 'Are there any instances of people having left your employ and gone into munitions works and come back to the boats because they did not like shore work?' He responded, 'Dozens of them . . . The workers on canals are . . . absolutely contented; and if they do leave that employment, they invariably want to come back again.'

This optimism was misplaced. They may have come back because they genuinely loved the lifestyle. But it may also have been that, being illiterate, and having no other training, they, like Captain Randle back in 1858, had little else in the way of opportunity. Once the chance to leave the canals was offered, it was in many cases taken. In reality, life on the cut was becoming more and more dislocated.

Many boatmen had fought in the war, and some had learned to drive lorries, the use of which had been given a very strong boost during the war years. Some now left the canals to go into road haulage, which required similar skills as those needed to work the boats (and similar powers of endurance).

Here was a pattern of behaviour that would increase over the next fifty years. While some boaters abhorred the world of dry land, others, having tasted its forbidden fruits, were attracted to an existence where you stayed in one place, saw your kids given an education and enjoyed all the mod cons of modern life. What's more, the economics of canal work made them less and less attractive. After a brief blip of prosperity the canals continued to decline after the war. In the slump of the 1930s life got even harder for those boaters who remained. More and more left the canals.

Lack of maintenance and improvements had left the canal infrastructure crumbling, and problems in traversing the canals, never mind the frustrating and interminable delays in loading and unloading, with what was by then primitive equipment, put even more pressure on the boat families, who were still, for the most part, paid by the trip. (Carrying companies, eventually, agreed to daily payments for when boaters in their employ were held up, for example, by ice. As for the 'Number Ones', any delays were, effectively, paid for by themselves.)

The Grand Union Canal's own carrying company, set up in 1934, was a particularly aggressive operator, taking the majority of the London to Birmingham trade, and the smaller independents increasingly had to rely on scraps – cargoes that weren't worth the candle for the bigger outfits. This often meant the dirtier loads, the most obnoxious of which was that carried by the refuse boats which travelled between Paddington and rubbish dumps in West Drayton, taking London's filth with them. These were family boats, and it was a campaign to stop children living in these utterly awful conditions that formed the spearhead of a renewed movement to take children off the boats altogether, beginning in the late 1920s.

Led by the trade unions, pressure was applied to bring in an eight-hour day, a six-day week, paid holidays and all the other rights enjoyed by workers on land. But even the boaters themselves scoffed at such fanciful ideas. They knew the realities of canal economics, and that luxuries such as paid holidays could never have been affordable. Certainly not by themselves, if they were self-employed.

A family takes tea on the boat, 1874

And without the long hours worked by the boating families, the system, already tottering, would fall over altogether. The canal companies were fully aware of this, and opposed all attempts – by politicians, trade unions and moralists – to stop children living on the boats. They knew that if the children left the boats, so would their mothers, and thus two-thirds of the workforce would be gone.

Missionaries, although they worked hard to better the conditions of life on the cut, were also opposed to measures to remove children from the canals, as to do so would fly in the face of their sincerely held beliefs concerning the sanctity of family life, which would thus be fractured.

However, the people who were most adamant that boating families should stay intact were the boaters themselves, or certainly the hard core among them. In the first place they quite naturally didn't want to be parted from their children. In the second, despite its hardships and cruelties, many loved the life they had led for generations and continued to lead even after they'd seen the delights offered by a life on land. Over 1,000 boaters signed a petition in opposition to a Private Member's Bill of 1929, aimed at taking children off the boats.

Nevertheless, because of outside pressure and an increasing desire on the part of some boating families that their children should be able to read and write, it became more common for boaters' children to live with a relative on land during their early years of schooling, in term time. But they often left while still at school age to rejoin their parents on board.

Despite the existence of local education authority 'kid catchers' on the canals (who could easily be evaded), it was still very difficult, even for families who actively wanted their children to go to school, to make it happen. Only when a boat was moored, perhaps waiting to be loaded or unloaded, to be given a new job, or for repairs, were children able to snatch more than a couple of hours' schooling. And then, at different schools. They were also regularly cold-shouldered by the children (and sometimes teachers) at the ordinary schools they infrequently attended, mocked and bullied because of their travelling lifestyle, lack of letters and their – perceived – dirtiness and lack of manners. And while many boaters did understand the benefits of education, others didn't. Some parents, especially fathers, resented the idea that their children might look down on them if they did learn their letters and become a 'scollard'. Some thought – and with good reason – that if their children went to school and were able to attain qualifications that led to greater opportunities they would leave the cut.

These pressures – the increasing desire for their children to receive a proper education, the decreasing prosperity, not to say the viability of the canal system – led to more and more boaters leaving the canals for good. By the end of the Second World War, only around 1,000 people were still living on the boats.

Literature and the Canals

In *Hard Times*, published in 1854, Charles Dickens provides one of his rare mentions of canals, and it's not a happy one. This is his description of the fictional 'Coketown' – a generic northern industrial town, perhaps inspired by Preston, with parts of Manchester, Cottonopolis itself, added in:

It was a town of machinery and tall chimneys, out of which interminable serpents of smoke trailed themselves for ever and ever, and never got uncoiled. It had a black canal in it, and a river that ran purple with ill-smelling dye, and vast piles of buildings full of windows where there was a rattling and a trembling all day long, and where the piston of the steam engine worked monotonously up and down, like the head of an elephant in a state of melancholy madness.

Eighty years later, George Orwell's *The Road to Wigan Pier* (the infamous pier was on the Leeds and Liverpool Canal), published in 1937, is writing in exactly the same vein: 'The train bore me away, through the monstrous scenery of slag-heaps, chimneys, piled scrap-iron, foul canals, paths of cindery mud criss-crossed by the prints of clogs.'

These grim depictions are one half of what became the standard trope about canals, a binary opposition that would occur, in slightly different but persistent forms, right down to our own day: the rural areas pierced by canals, good; the industrial areas, bad. (And the post-industrial world of high-rise estates, worse still.)

With the exception of passenger trips (for example, during the annual 'Wakes Weeks', when industrial workers were given a holiday), leisure boating was rare on the inland waterways. Rivers began to become popular with pleasure craft in the last quarter of the nineteenth century, canals far less so.

In 1868 came an interesting exception to this rule. A

plucky pioneer, J. B. Dashwood, recorded a journey, largely by canal, taken on a small sailing boat. The *Caprice* travelled from the Thames at Weybridge down to the Solent at Littlehampton on the Wey and Arun Canal, which had been finished in 1816. The trip, undertaken by 'your humble servant and his better half' (who must have been the silent type, as she is very rarely mentioned again), along with a dog, Buz, and a portable India-rubber bath, sets the tone for scores of books that were to follow. A well-to-do chap and his wife enjoy the scenery and wildlife on a leisurely pootle-about along a waterway, stopping to admire old priories, Tudor manor houses, ruined palaces and castles en route (with much erudite history thrown in). Their enjoyment tempered only by the sorry state the canal itself had been allowed to fall into.

A canal community in London, late nineteenth century

For Dashwood gives us a valuable insight into the parlous state of many 'agricultural canals', even by this early date. Weeds have been allowed to grow in abundance, fouling the propellers of a steamer. The lock machinery is stiff. Local landowners have placed a series of gates along the towpath at regular intervals, some of which are so elaborate as to mean that the horse has to be completely uncoupled to pass through. Water has been diverted by millers and farmers: at one point one of them has to be bribed with half a crown so as to let the *Caprice* continue on its journey. The canal is 'almost at a standstill from want of use'. Worse is to come. A branch to Chichester that Dashwood had intended to take has been out of commission, he finds out, for the past eleven years, trodden in by cattle, or filled in by man, and bone dry.

Of the Wey and Arun Canal itself he concludes:

> *How long this connecting link will be available seems most doubtful, for the whole of this Canal is, I understand, to be put up immediately for auction. In fact, considering the great expense in procuring water, the competition of the railway, and the small amount of traffic, it is impossible it can pay its way. It will be a sad pity if it ceases to exist, for the scenery after leaving Bramley is most lovely, and we thoroughly enjoyed our trip.*

It was abandoned three years later, but here are the first seeds of an attitude that would lead to a mighty movement to save and restore these waterways in our own time.

(One notable, if unsuccessful, attempt to 'save' an ailing agricultural canal – the Basingstoke – was undertaken by the unusual means of wholesale fraud. This was the brainchild of an extraordinary individual, Horatio Bottomley, whose portfolio career combined the jobs of a serving MP and editor of the jingoistic magazine *John Bull* with that of professional swindler. Along with some cronies, Bottomley took over the Basingstoke Canal, in the guise of the grand-sounding 'London and South Western Canal Company'. The name cunningly fashioned so as to suggest a connection with the London and South Western *Railway*, a business relationship that did not in fact exist. Thousands of duplicated shares were issued, and Bottomley was extremely lucky to escape conviction when angry shareholders discovered their shares were worthless. He was finally brought to book and imprisoned for an unrelated caper in 1922. A prison visitor, to his great satisfaction, came across this once-mighty figure engaged in the production of mail bags. 'Sewing, Bottomley?' 'Reaping,' came the reply.)

It wasn't until the turn of the nineteenth century that more books about the waterways began to trickle out. George Westall's *Inland Cruising on the Rivers and Canals of England and Wales*, published in 1908, was a practical guide as to the conditions of various canals and navigations, as well as providing information about their history. After the First World War Westall became president of the National Inland Navigation League, which promoted the benefits of the waterways.

In 1911 E. Temple Thurston published *The Flower of Gloster*, which purports to be a journey from Oxford to Lechlade, via various canals. Doubts remain as to how much of this journey the author actually undertook. Large sections of the book deal with the more civilized delights of the surrounding towns, while the boat's skipper,

'Eynsham Harry', is patronized throughout as a kind of idiot savant, whose homespun wisdom is reported with the same indulgent clucking as might accompany the pronouncements of a particularly precocious small child.

As with his predecessor, Captain Randle, Harry's disquiet about his inability to read is laughingly dismissed by the author. No such person as 'Eynsham Harry' is known to have existed, and while Thurston may well have constructed a composite character from more than one real boatman, the suspicion remains that the author – a prolific popular novelist and playwright – simply made him up. Of undoubted interest is the passage where Thurston and Harry leg their way through the Sapperton Tunnel – some of the last people to do so, as it closed in 1911.

In 1916 what amounts to an early form of blog was published by P. Bonthron, *My Holidays on Inland Waterways*. The author visits some fourteen canals and has nothing of interest to say about any of them.

In 1930 A. P. Herbert published his novel *The Water Gipsies*, which includes a trip along the Grand Union Canal. Unlike Temple Thurston the author – a famous one in his day – takes the trouble to do some research before putting pen to paper. (Unlike Thurston, he doesn't refer to a narrow boat as a barge. Although he does persist with the mythology Thurston created: that the boaters have Gipsy origins.) The book is broadly sympathetic to the boaters – although the illiteracy of her boatman suitor becomes an insuperable problem for the novel's heroine.

Herbert provides several fascinating glimpses of the boat families and their floating homes. The boatmen's wives are *queens of these little kingdoms, dark-eyed, gipsy-looking, weather-worn creatures, with shawls over their heads and children crawling under their feet, who stood at the doors of their cabins as they have stood for generations, patiently waiting for the next move forward in their wandering existence . . . The tiny cabin before them, a few feet square, and not high enough to stand up in, held all their worldly goods, their kitchens, their romances, their bridal-beds, their husbands and babies, their past and present and all their future.*

The boats' hatches are painted with the distinctive designs of roses, castles and hearts. Inside the cabin were

smaller castles and smaller hearts and panels of diamonds and clusters of roses, in brilliant reds and yellows and greens; and every domestic vessel – the great water-can on the deck, the biscuit-tin on the shelf – had its roses. Just inside every door on the left-hand side, in exactly the same place, were fixed a few shining brass knobs and bosses, like the ornaments of bedsteads and cart-horses; and, highly polished, they made a brave show . . . Beyond the brass knobs was the tiny stove, polished and speckless. On the stove stood always a vase of wild flowers . . . and on each side of the stove were hung festoons of ornamental plates – fine decorated plates with gilded edges or filigree borders, and pretty pictures of Victorian ladies or dancing shepherdesses, and in great gold lettering, 'A Present from Bombay' or 'Banbury Cross'.

Herbert also describes the elaborate and very distinctive rope work on the ram's head on top of the rudder. Often, when a faithful horse died, its tail and a scrap of hide were tacked on too, so that the spirit of the animal could continue to smile on the boat family's fortunes. Strangely, Herbert doesn't mention the very distinctive black bonnets, rather like shorter versions of Spanish mantillas, worn by the boatwomen. They were originally white but turned to black in 1901, in mourning for Queen Victoria. Only in the 1930s were they to revert to happier colours, before disappearing altogether.

In 1933 the canoeing enthusiast William Bliss published *The Heart of England by Waterway*, an account of several journeys he and some companions undertook, beginning in the late 1890s. They too pass through the Sapperton Tunnel. Among many lyrical passages describing the flora and fauna and rural scenery, there is a *cri de cœur* that could have come verbatim from books written in the 1940s, 50s or 60s.

> *One by one our English Canals are becoming derelict; each lustre sees another go. Once upon a time, not so very many years ago, it was possible to travel all over England, north, south, east and west, by river and canal; there was not a county you could not visit, hardly a town you could not reach by water, if you liked and if you were not (and what lover of boats and rivers ever was or will be?) in any particular hurry to get there ... They are going one by one, and I am writing thus about them before they go, so that anybody who reads and who has never yet experienced the peaceful beauty, the sleepy contentment that is peculiar to these English waterways, may do so before it is too late.*

It was too late.

The Second World War

Canals gained another wind during the Second World War. Although some that had already run out of traffic, such as the Kennet and Avon, Basingstoke, and Wey and Arun Canals, were used to make fortifications in case of an invasion, others were put to active use aiding the war effort. A propaganda film-cum-unconvincing love story made in 1944 by Ealing Studios, *Painted Boats* (alas it was in black and white), with a commentary by the poet Louis MacNiece, made a case for the vital role canals played in the transport network – and could do in the future – by taking heavy traffic off the roads and railways.

Regent's Canal, 1944

We must not overlook the cut itself or any of the people who work on it, by it or for it. Their work is part of the war effort, and should be, later, part of our peace effort. The work goes on. There's life in the old cut yet.

It seemed, for one glorious moment, as if the government actually saw a commercial future for the canals. Once again, it was a false dawn.

One development during the war years was, however, to lead in our time to the movement that has done so much to revivify the canals. This was the publication of a book that, almost single-handedly, sparked interest in the now half-forgotten and more than half-neglected waterways. It was *Narrow Boat*, written by L. T. C. Rolt, and published in 1944. The book tells the story of a journey that Rolt, a mechanical engineer, had taken with his wife in 1939 along various Midlands canals. Like Evelyn Waugh's *Brideshead Revisited*, also written while Britain was at war, the book is a lament for a world both authors feared would be destroyed: the world of aristocratic privilege in Waugh's case, 'the spirit of an older and happier rural past' in Rolt's. In his autobiography, published in 1977, Rolt admitted that at the time he wrote the book canals 'represented the equivalent of some uncharted, arcadian island inhabited by simple, friendly and unselfconscious natives where I could free myself from all that I found so uncongenial in the modern world'. (*Brideshead Revisited*'s first chapter is called 'Et in arcadia ego'.)

That the book conjured up a powerful and attractive (and sincerely meant) arcadian fantasy there is no question. Nor is there any doubting Rolt's vehement hatred of every aspect of the modernity that he found so uncongenial. Council estates, commercial travellers, cinemas, cafés, processed cheese, even dartboards are routinely described as garish, hideous, tawdry, drab, dreary, squalid, dingy and mass-produced – or a combination of several of those epithets.

The opinions expressed in *Narrow Boat* are at the heart of an attitude that has continued to characterize our view of canals to this day: the canal as a rural retreat, a refuge from the unpleasantness of modern life (which the inland navigations, as handmaidens to the industrial revolution, had done so much to create).

Such was the great success of *Narrow Boat* (entirely unexpected by its author) that interest in the canals was reawakened. This included interest in their history. Thanks to nationalization, which brought the papers of the canal companies under one roof, at the Public Records Office, a serious attempt to write a modern account of them was now possible. The main figure involved in this was Charles Hadfield, who, using company records and newspaper reports from the time, constructed an in-depth history of the canal system in a whole series of scholarly books he either wrote or edited.

In 1946 Rolt and Charles Hadfield were approached by a literary agent and author with an interest in canals, Robert Aickman. At his urging they formed the Inland Waterways Association. This was what we'd now call a pressure group, lobbying for canals and navigations to be saved from destruction.

It was a fractious outfit, the high hand of the pugnacious Aickman soon turfing both Rolt and Hadfield out of the nest. They had disagreed on strategy: while Aickman wanted to fight tooth and nail to save any and every waterway under threat, the others felt it more politic to target particularly vandalistic schemes to close canals down, not believing it possible, or practical (or perhaps even desirable), to save each and every one. The early

IWA had little interest in the less romantic big canals, which remained commercially successful.

Beginning in the late 1940s, several books appeared that gave a real insight into life on the cut, written by people who had worked as boatmen – and boatwomen. The first, published in 1948, was *Maidens' Trip*, written by Emma Smith, one of the so-called 'Idle Women'. Their nickname came from the initials (which stood, of course, for 'Inland Waterways') on the badges they had worn when, from 1942, they were invited to do their bit for the war effort by working as boat crews for the Grand Union Canal Carrying Company (and, later, the Ministry of War). Like the 'Land Girls' and the women who drove trucks and ferried aircraft up and down the country, they were far from idle, doing a tough and important job. Quite quickly, after some initial prejudice, they won the respect and friendship of the great majority of the boaters they encountered.

Though there were never more than eleven pairs of boats worked by these volunteers at any one time, the Grand Union scheme proved to be, for female writers, every bit as fertile as the New York Actors Studio was to become for 'Method' actors. *Maidens' Trip*, a fictional account of Emma Smith's time on the cut, won the prestigious James Tait Black Memorial Prize. But it was only one of four books that were published by former GUC inmates. Susan Wolfitt – wife of actor Sir Donald Wolfit (he dropped one 't' in his stage name) – wrote *Idle Women*; Margaret Cornish, *Troubled Waters*; and Eily Gayford, *The Amateur Boatwomen*. 'Kit' Gayford, a former ballet dancer, had spent three years training the women volunteers.

All of these books are an invaluable source for the social history of the canals. They describe the sheer slog and danger of canal boat work: cracking your skull against an oncoming bridge if you weren't looking, falling into the lock and being crushed between the boat and the gates or walls, scraping your knuckles raw if you held the tiller too near a tunnel wall. Steering a canal boat, never mind jumping on to it from the lockside in wet weather, wasn't as easy as it looked.

A woman training to work on a narrow boat, 1944

The boaters had a lingo all of their own, and woe betide anyone using the terminology of the Senior Service. Canals were 'cuts'. Narrow boats were never, ever to be called 'barges'. Nor was a person working such a vessel to be referred to as 'bargee' – except by ignoramuses 'off the land'. It wasn't then and isn't now the done thing to peer into a cabin. If you want to cross from one boat to another you call out and ask if it's all right to do so and then cross at the fore-end. To break such rules was to run the risk of being regarded as what de Salis called a 'gongoozler', but which the boaters were more likely to call a 'Rodney', an all-purpose term for any canal pariah.

There was no port or starboard. A steering instruction would be to 'hold in' or 'hold out', depending on which side the towpath was. When you got underway you didn't 'cast off', you 'let go'. The bows were the 'fore-end', ropes were 'straps' or lines. The rope connecting the motor boat to the butty was a 'snubber'. A boat hook was a 'shaft'. Sluices were 'paddles', operated by a windlass. 'Lock-wheeling' was the job of racing down to the next lock on a bicycle to get it ready for the boat.

An especially interesting insight into the lives – and feelings – of the boatwomen, and their children, was given in an outstanding piece of oral history, Sheila Stewart's *Ramlin Rose*. Published in 1993, it's made up of interviews with several retired boatwomen about their lives on the cut from the First World War up until the 1960s. The heroine of A. P. Herbert's *The Water Gipsies* had found the boatwomen mysterious, taciturn and somewhat frightening. Here they reveal themselves, in living, and very moving, colours.

The wife of a boatman was known as his 'Best-Mate'; he was sometimes known to her as 'Moy-Chap'. The men, as is often the way with such creatures, weren't always expansive in acknowledging the contribution their wives made to the family business. Sheila Stewart quotes one wonderful exception to the rule. When an outsider, the skipper of a Birmingham 'joey', or day-boat, complains about 'that woman', who is steering the family boat, and delivers himself of the opinion that women shouldn't be allowed to do so, 'Moy-Chap' gives him both barrels.

> *That 'woman' is my Best-Mate. She's a better boat man than you'll ever be, with yer filthy boat and yer knackered 'orse. She's stroved with boats all her life. She's stroved with boats all through the bloody war . . . if it wasn't for the strovin these women, and the women before them, and the women before that, their grandmother and great-grandmother, my grandmother and great-grandmother, all unpaid labour, the canals would of died out years ago; there'd be no bloody Cut, and you'd be out of a job, Mate!*

Emma Smith was the first of a number of middle-class visitors to professional boating life who reported on its perils and pitfalls and, unlike the boaters, who were inured to them, made her frustrations known. She, like the other writers who followed, was flabbergasted by the sheer inefficiency and wastage caused by the dilapidated state of the wharves. Not only could it take days until a cargo was unloaded, she watched antique loading gear pour as much of her shipment of coal into the canal as it did into the boat.

The same sense of sheer exasperation at the dreadful

state of the canals' infrastructure and working practices was expressed in *Hold On a Minute*, published in 1965, by Tim Wilkinson, a former engineer who had been wounded in the war. It was based on a year he had spent as a captain of a pair of working narrow boats in 1948 with his wife Gay, a former model and film actress. (Canals seem to have held a fatal attraction for actresses, models and dancers. Sonia South, another actress who worked with the 'Idle Women', went on to marry a working boatman and lived on the cut for several years, before becoming the wife of Tom Rolt and a vice president of the Inland Waterways Association.)

One shocking aspect that emerges in these books is the hostility with which boaters were routinely treated by the non-floating population. They were spat on from bridges and had stones and other missiles thrown at them by wastrels on the banks. They were often refused service in shops.

A skipper tells Tim Wilkinson that his wife and daughter

have sometimes been told to get out. 'No gipsies served in here,' say some shops. It hurts, you know. We are poor people but proud, and anyway, after being insulted so often, we kind of go into our shells, if you follow me. We don't stick our neck out asking to be beat.

There is no doubt about the great affection – Wilkinson describes it as love – that he and other authors had for the remarkable men and women they encountered on the canals. Taken into their trust, they gained an important and instructive insight into life on the cut. (Not least, the court-ship rituals of young couples who may only encounter each other once every several weeks – they declared their affections by chalking their initials under bridges.)

The problem with this tendency to canonize the boaters is that not only are they often portrayed as childlike, naive and superstitious, but their very lack of knowledge of the modern world comes to be seen as a precious commodity, a living link to Rolt's rural arcadia. As he put it in *Narrow Boat*:

> *The canal . . . has become the last remaining stronghold of a people whose way of life has survived the whole course of the revolution substantially unchanged, and who therefore retain to this day many of the characteristics of the pre-machine age peasant.*
>
> *The men I have been privileged to meet . . . revealed more eloquently than any words of mine a way of life, which was the antithesis of the stereotyped and root-less existence of twentieth-century 'economic man', and it is the spirit of which these men are the unthinking guardians which must not be permitted to perish from the earth.*

And things were changing. Although canal boatmen were classified as working in a reserved occupation, and so not called up during the Second World War, National Service now claimed the young men. Having tasted life on land, with all its creature comforts, many decided to leave the cut. This development horrified the Roltian romantics, who came to see the continued existence of the boating families as a kind of moral and social barometer. As long as they remained in harness, there was yet some hope for the world, and the dream of halting the march of modernity stayed alive. The 'do-gooders', as Wilkinson describes them several times in his pages, who wanted to tempt the families off the narrow boats, came in for special scorn.

As before, literacy was the nub of the issue. With great reluctance, Tim Wilkinson agrees to teach a young boater who joins him as mate how to read and write, at the young man's own request. He wants to do so, he says, so he

> *could go on the land and get a proper job like some of me mates. The Army teached they, and they never come back . . . Them as has gone talks to the women about cinemas, dancing and all what goes on. All the women wants to leave the boats.*

Contemplating the contribution made to this growing exodus by National Service, Wilkinson reflects that

> *The boys from the cut were taught to read and write, to drive lorries, to clean their teeth, to become dissatisfied with their life on the canal. But after their discharge from the Services, what? An unskilled job, a street corner, a cheap dance-hall floor, a steamed up 'caff' to spend their evenings in, a sordid room in a dirty lodging house . . . I felt deeply for them, and could appreciate their reaction as the novelty wore off and the realization of their tragic mistake dawned.*

That's not to say that Wilkinson is entirely wrong, and that life on land, with its prospects of only unskilled and monotonous labour, was a paradise come true. Talking of boatmen who had left the cut to work in factories, a captain tells him, 'Take us off the boats, and we wither like a picked flower.'

But working life on the cut was doomed. A wartime

committee formed to assess the future of the inland waterways had already come to the grim conclusion that 'for all canals with locks capable of holding only one narrow boat at a locking the prospect is bleak and unpromising and should be faced. They are definitely uneconomical and their financial position is unsound.'

In 1948 Fellows, Morton and Clayton, by far the biggest carrying company and an important bell-wether for the health of the whole canal system, declared a loss for the first time since it had set up shop in 1889. The same year it went into voluntary liquidation. This could hardly be seen as an encouraging sign.

And things got worse. In 1949 the Docks and Inland Waterway Executive, part of the British Transport Commission, declared of its remit: 'the future of the artificial waterways is obviously considered doubtful'. (As opposed to that of the docks.) It favoured making improvements only to those navigations and canals that connected to the estuaries. The DIWE, though, did float the idea of pleasure boating on some canals, including the Llangollen. The BTC refused to countenance such a preposterous development.

The sad and sorry decline of large parts of the canal network thus continued. Even the route to and from Birmingham and London, one of the lifelines of the whole system, suffered from neglect and indifference.

In 1984 a memoir called *Bread Upon the Waters* provided a snapshot of the canals in the early 1960s. Like Tim Wilkinson, David Blagrove was a refugee from the rat race who spent time as a professional boatman, working as the captain of a pair of narrow boats on the London to

Birmingham route in 1962 and 1963. The book grimly demonstrates that, in the fifteen years since Wilkinson had been on the canals, the neglect of the network had become even more pronounced under the wayward stewardship of the British Transport Commission. Like Wilkinson, Blagrove leaves the cut after a year, as the grindingly hard life of the professional boater simply doesn't pay.

He despairingly describes 'the background of frustration, incompetence and apathy which so nearly lost us our canal system'. The lack of maintenance, of both canals and reservoirs, the lack of dredging, the lack of water – all make the canals more and more impassable, and impossible. The lack of up-to-date facilities at depots and wharves (and, according to Blagrove, union intransigence) make loading and unloading an even more agonizing and wasteful process than it had been when Emma Smith and Tim Wilkinson saw its many shortcomings at first hand in the 1940s.

Bread Upon the Waters also describes the noisome pollution being discharged into the rivers and canals by factories – which the canal authorities allow, and indeed take a fee for. The 'run-off' from road surfaces adds bitumen to this already toxic mix. Dead dogs, prams, tyres, mattresses (and in later years beer cans, shopping trolleys and polystyrene packaging) all find their way into the water. Barbed wire is helpfully thrown in so that it can snarl up propellers.

Blagrove gets caught up in the famously terrible winter of 1962/3, and his boat is stuck in ice for more than two months. This disastrous season of discontent – on the part of the customers as well as the boaters – has gone down in legend as the final nail in the coffin for the canals, one of the main reasons everyone associated with their commercial traffic threw in the towel. It's not quite the case but it certainly didn't help the cause.

The main problem was that the BTC simply didn't believe in the future of the inland waterways. Indeed it was, in some cases – such as that of the Kennet and Avon – actively trying to hasten their demise by deliberately rendering whole sections unnavigable. In the early 1950s the BTC tried to hand the canals, over lock, stock and barrel, to local authorities – which the latter, knowing a 'hospital pass' when they saw one, refused to agree to.

All over the country plans were being made to drain, fill in and pave over canals. One argument – and because of the rapidly deteriorating state of many urban canals it had some merit – was that they were dangerous, causing children to be injured or drowned. Many canals were decommissioned. (While some grudging improvements were made to those the BTC considered viable.) But now, with the canals facing the final curtain, the tide began, very slowly, to turn. The attempt on the part of the BTC to get rid of them altogether had rallied the conservationist troops, and the movement to keep canals open grew in strength from that moment on.

Though some local authorities were sympathetic, and while the National Trust was persuaded to take over the running of some waterways, this campaign – in its way as remarkable as the early days when canals were first built – came from the grass roots, fought by individual men and women determined to save the canals from oblivion.

The Inland Waterways Association inspired and supported local canal trusts that lobbied for their continued existence. Volunteers made the all-important voyages – even if it was only in canoes – that registered the fact that there was indeed demand for the canals and navigations and stopped the BTC from closing them down. The Kennet and Avon was a case in point.

Even more extraordinary was the great groundswell of effort physically to restore the derelict sections of canals, which is still going strong today. This was – and still is – achieved, very largely, by an army of amateur enthusiasts, in many cases equipped, like their forebears, with little more than a wheelbarrow, pick, shovel and paintbrush.

When this movement first began, in 1949, navigations were its target. The first of many success stories on the canals per se was the restoration of the then unnavigable southern section of the Stratford-upon-Avon Canal. Locks were rebuilt, and the canal reopened in 1962, inspiring other such remedial work elsewhere. (Notably 'Operation Ashton', on the Ashton Canal, in the late 1960s.) More and more miles of half-forgotten waterways were brought back from the dead – at the present time running into a total of several hundred miles of them. Warehouses, lock cottages and canal pubs – many of them fallen into disrepair and subject to vandalism – were also restored. (In 1970 a senseless piece of arson had destroyed Telford's fine warehouses at Ellesmere Port.)

The main reason the canals are still alive and well and with us today, however, is unquestionably because of their having been converted into waterways for leisure. In the late 1940s pleasure craft on the canals were rare, and their presence discouraged by both canal authorities and boaters, for obvious reasons. But from the early 1950s the leisure trade on the canals began to mushroom. The annual Festival of Boats, run by the IWA, began in 1950, initially, at Market Harborough, and is still going strong. Boat hire firms began to spring up from 1952.

This increased emphasis on pleasure boating wasn't to everyone's taste. In 1948 a working boatman had told Tim Wilkinson: 'We're doomed, I reckon, and in a few years what's left of the cut will be like a kid's pond – just covered with a lot of toy boats running around amusing themselves.'

In its early days the IWA itself bent its efforts to supporting the commercial trade rather than the leisure business, but it was fighting a losing battle. By the late 1960s practically all commercial traffic on the narrow network had ceased. Coal was rapidly being replaced by oil as the fuel of choice for the factories that had moved to the canalsides to gain access to it. Only power and gas stations prolonged the carriage of coal, until they too began to switch to oil. Only the broader canals serving estuaries kept afloat.

As L. T. C. Rolt put it in 1969, 'A map of the waterways that are still used commercially today has virtually contracted to the shape it was before the Duke of Bridgewater, Gilbert and Brindley embarked upon their daring enterprise. The wheel has come full circle.'

Meanwhile, as commercial traffic drained away, leisure boats took over. In the course of the 1950s the number of cruising licences issued increased from 1,500 to over 10,000.

By now it was clear to anyone, friend or foe, that the only future for the canals was to be that of aquatic pleasure grounds. One body which certainly thought so was the British Waterways Board, which had succeeded the hated British Transport Committee in 1963. The BWB was to prove far more sympathetic to the canals than its slash-and-burn predecessor and actively promoted the leisure trade – not least, in a series of stultifyingly tedious promotional films.

Despite, if not because of these stilted productions, leisure traffic on the canals increased hugely throughout the 1960s and 70s, with more and more hire boats and private pleasure craft afloat (many adapted from working boats). Of course, this could only happen because steps had been taken to maintain the locks and other infrastructure, to limit if not prevent pollution, certainly in the tourist areas, and, rather obviously, to make sure there was water in the canals. A privilege that hadn't always been afforded to the last working boaters who had used them.

Today it is generally reckoned that there are more boats on the inland waterways than there were at the height of their industrial heyday.

It isn't just pleasure craft that benefitted from this renaissance. Towpaths, some of which had been off-limits to outsiders when the commercial trade was still going, opened up to walkers, and later cyclists. Anglers had always used the canals, and continue to do so. Many canoeists paddle up and down the waterways, and on the reservoirs all manner of sailing and windsurfing clubs have sprung up to take advantage of these amenities.

Canals – partly thanks to their long period of virtual

abandonment – had become homes to much rare flora and fauna, as well as being haunts of coot and tern, grebes and kingfishers. Wildlife lovers are today offered a rich source of study.

Docks too, having fallen into commercial disuse, have been spruced up, at Liverpool's historic Albert Docks, the Clydeside, Gloucester, and Salford Docks and Bristol's famous Floating Harbour. Once paved-in canals – like the Thames and Severn at Stroud – are now being reclaimed for leisure use. Apartment blocks have been built on the sites of the former basins, depots and wharves. When these developments began in the 1960s and 70s, many old buildings were wantonly destroyed, until their value as attractive shells for residential and leisure development finally came to be appreciated. Many are now sensitively adapted to new use.

Around 15,000 people now live on boats moored (sometimes permanently) on our inland waterways. There are several museums, great and small, on today's canal network, some of them 'living museums', such as the one at the archetypal canal village at Stoke Bruerne on the Grand Union Canal.

The exploration of canal history has blossomed, especially in the field of oral record. Interviews with the last surviving boat family members, and others who worked on the canals, have helped fill in a fascinating picture of the social history of the men, women and children who were born, worked, courted, married and died on the cut.

Today the trickle of articles and books about canals, which began in the 1850s, has become a flood, with dozens of works on canal history and accounts of canal cruises cascading on to the bookstalls each year, plus the many blogs, forums and 'virtual tours' online. In 2012 the Canal and River Trust took over from the British Waterways Board. It too has a strong presence on the web.

Two hundred and fifty years after two remarkable individuals cut the first sod of the 'Grand Trunk', the canal system they did so much to bring into being, having suffered a near-death experience, is alive and well and looking forward to the next chapter.

Further Reading

General History

Joseph Boughey, Charles Hadfield, *British Canals – The Standard History* (2008). The updated definitive account, in one volume.

Anthony Burton, *The Canal Builders* (1972). A well-written account.

Charles Hadfield, *Thomas Telford's Temptation* (1993). The book which set the Telford/Jessop hare running.

Peter Lead, *Agents of Revolution* (1989). It deals with the important contribution of John and Thomas Gilbert to the early canal age.

Christopher Lewis, *The Canal Pioneers – Brindley's School of Engineers* (2011). A volume that sheds important light on Brindley's assistants.

E. Paget-Tomlinson, *The Illustrated History of Canal and River Navigations* (1993). An indispensable textbook covering every aspect of inland waterways.

L. T. C. Rolt, *Navigable Waterways* (1969). The most readable single-volume history, unsurpassed.

The Eight Canals

The cornerstone of canal history was put into place in a series of books written or edited by Charles Hadfield. The works consulted for this volume are:

Charles Hadfield, *The Canals of the East Midlands* (1966)

Charles Hadfield, *The Canals of the West Midlands* (1966)

Charles Hadfield, *The Canals of South and South East England* (1969)

Gordon Biddle, Charles Hadfield, *The Canals of North West England* (1970)

Charles Hadfield, *The Canals of Yorkshire and North East England* (1972–3)

Plus

S. R. Broadbridge, *The Birmingham Canal Navigations* (1974)

A. D. Cameron, *The Caledonian Canal* (1994)

Mike Clarke, *The Leeds and Liverpool Canal* (1990)

Kenneth R. Clew, *The Kennet and Avon Canal – An Illustrated History* (1968)

Alan H. Faulkner, *The Grand Junction Canal* (1993)

Peter Lead, *The Trent and Mersey Canal* (1980)

Len Paterson, *From Sea to Sea – A History of the Scottish Lowland and Highland Canals* (2006)

Ray Shill, *Birmingham and the Black Country's Canalside Industries* (2005)

Mike Taylor, *The Canal & River Sections of the Aire & Calder Navigation* (2003)

Historical Texts

William Bliss, *The Heart of England by Waterway* (1933)

J. B. Dashwood, *The Thames to the Solent – By Canal and Sea* (1868)

John Hassell, *Tour of the Grand Junction Navigation* (1819)

A. P. Herbert, *The Water Gipsies* (1930)

John Hollingshead, *On The Canal* (1858)

John Phillips, *A General History of Inland Navigation* (1805)

Joseph Priestley, *Historical Account of the Navigable Rivers, Canals, and Railways of Great Britain* (1831)

Henry Rodolph de Salis, *Bradshaw's Canals and Navigable Rivers of England* (1904)

Samuel Smiles, *Lives of the Engineers* (1862)

E. Temple Thurston, *The Flower of Gloster* (1911)

George Westall, *Inland Cruising on the Rivers and Canals of England and Wales* (1908)

Social History

The most important works in the field are:

Wendy Freer, *Women and Children of the Cut* (1995)

Harry Hanson, *The Canal Boatmen 1760–1914* (1975)

Harry Hanson, *Canal People* (1978)

Many interesting books of oral history have been written in recent years, the most outstanding being Sheila Stewart, *Ramlin Rose: The Boatwoman's Story* (1993).

Post-1940 Novels and Memoirs

David Blagrove, *Bread Upon the Waters* (1984)

Margaret Cornish, *Troubled Waters* (1987)

Tom Foxon, *Number One!* (1991)

Eily Gayford, *The Amateur Boatwomen* (1973)

L. T. C. Rolt, *Narrow Boat* (1944)

Emma Smith, *Maidens' Trip* (1948)

Tim Wilkinson, *Hold On a Minute* (1965)

Susan Wolfitt, *Idle Women* (1947)

Gazetteers and Guides

Robert Aickman, *Know Your Waterways* (1955). This is a commandment, not an invitation.

Anthony Burton, Derek Pratt, *The Anatomy of Canals* (2001–4). The first two volumes include good accounts of the topography, as well as the history of the major canals.

Collins Nicholson Waterways Guides. A series with solid practical and geographical information.

Stuart Fisher, *The Canals of Britain: A Comprehensive Guide* (2012). Contains much historical information.

Hugh McKnight, *The Shell Book of Inland Waterways* (1975). Particularly strong on architecture and history. Sadly, not revised since 1978, but still useful.

Michael Pearson, *Pearson's Canal Companions*. A lively, informative and erudite set of guides.

Image Credits
Historical images

Additional photography

Adrian Ashworth: Trent and Mersey Canal, Birmingham Navigations, Aire and Calder Navigation, Leeds and Liverpool Canal, and Grand Union Canal
Anthony Macmillan: Caledonian Canal
Anthony Brown: Kennet and Avon Canal

Maps by Jeff Edwards

He just wanted a decent book to read ...

Not too much to ask, is it? It was in 1935 when Allen Lane, Managing Director of Bodley Head Publishers, stood on a platform at Exeter railway station looking for something good to read on his journey back to London. His choice was limited to popular magazines and poor-quality paperbacks – the same choice faced every day by the vast majority of readers, few of whom could afford hardbacks. Lane's disappointment and subsequent anger at the range of books generally available led him to found a company – and change the world.

'We believed in the existence in this country of a vast reading public for intelligent books at a low price, and staked everything on it'
Sir Allen Lane, 1902–1970, founder of Penguin Books

The quality paperback had arrived – and not just in bookshops. Lane was adamant that his Penguins should appear in chain stores and tobacconists, and should cost no more than a packet of cigarettes.

Reading habits (and cigarette prices) have changed since 1935, but Penguin still believes in publishing the best books for everybody to enjoy. We still believe that good design costs no more than bad design, and we still believe that quality books published passionately and responsibly make the world a better place.

So wherever you see the little bird – whether it's on a piece of prize-winning literary fiction or a celebrity autobiography, political tour de force or historical masterpiece, a serial-killer thriller, reference book, world classic or a piece of pure escapism – you can bet that it represents the very best that the genre has to offer.

Whatever you like to read – trust Penguin.